LAUNCHERS, LOBBERS, AND ROCKETS
ENGINEER

Brimming with creative inspiration, how-to projects, and useful information to enrich your everyday life, Quarto Knows is a favorite destination for those pursuing their interests and passions. Visit our site and dig deeper with our books into your area of interest: Quarto Creates, Quarto Cooks, Quarto Homes, Quarto Lives, Quarto Drives, Quarto Explores, Quarto Gifts, or Quarto Kids.

© 2018 Quarto Publishing Group USA Inc.

First published in 2018 by Rockport Publishers, an imprint of The Quarto Group, 100 Cummings Center, Suite 265-D, Beverly, MA 01915, USA.
T (978) 282-9590 F (978) 283-2742 QuartoKnows.com

All rights reserved. No part of this book may be reproduced in any form without written permission of the copyright owners. All images in this book have been reproduced with the knowledge and prior consent of the artists concerned, and no responsibility is accepted by producer, publisher, or printer for any infringement of copyright or otherwise, arising from the contents of this publication. Every effort has been made to ensure that credits accurately comply with information supplied. We apologize for any inaccuracies that may have occurred and will resolve inaccurate or missing information in a subsequent reprinting of the book.

Rockport Publishers titles are also available at discount for retail, wholesale, promotional, and bulk purchase. For details, contact the Special Sales Manager by email at specialsales@quarto.com or by mail at The Quarto Group, Attn: Special Sales Manager, 401 Second Avenue North, Suite 310, Minneapolis, MN 55401, USA.

10 9 8 7 6 5 4 3 2 1

ISBN: 978-1-63159-427-4

Library of Congress Cataloging-in-Publication Data is available

Design and page layout: Laia Albaladejo
Photography: Lance Akiyama and Kaile Akiyama

Printed in China

LAUNCHERS, LOBBERS, AND ROCKETS ENGINEER

MAKE 20 AWESOME BALLISTIC BLASTERS ⇨ with ordinary stuff

LANCE AKIYAMA

CONTENTS

- 6   INTRODUCTION
- 8   GOOD, SAFE FUN

10   SIMPLE AND SUCCESSFUL

- 12   Straw Blowgun
- 16   Micropult
- 18   Rubber Band Rockets
- 22   Pocket Bow

26   BOWS AND SLINGSHOTS

- 28   Pulley-Powered PVC Bow & Arrow
- 38   Slingshot and Arrow
- 44   Duct Tape & PVC Crossbow
- 50   Wrist-Mounted Crossbow

56 MINI MEDIEVAL SIEGE MACHINES

- 58 Onager
- 66 Da Vinci Catapult
- 72 Ballista

80 CURIOUS CONTRAPTIONS

- 82 Roller-Amplified Many-Thing Shooter
- 88 Desk Drawer Booby Trap
- 94 Poker Chip Shooter
- 100 Slide-Action Rubber Band Gun

104 FIREARMS

- 106 BBQ Blaster
- 112 Ping-Pong Ball Mortar
- 118 Soda Bottle Bombard
- 122 Handheld Rocket Launcher
- 134 Ballistics Gel Target

- 138 Resources
- 139 Acknowledgments
- 140 About the Author
- 141 Index

INTRODUCTION

Create, aim, and fire!

Greetings, aspiring blaster crafter! In your hands is a complete compendium of instructions to create your own crafty cache of DIY shooters!

Look no farther than your kitchen drawer and hardware store to find everything you need. The launchers and lobbers in this book are built from a few plastic bottles, some pipe, a little woodcraft and wire, and a whole lotta hot glue. The rocket fuel ranges from rubber bands to hair spray. You don't need to be a rocket scientist to get started, but you might feel like one by the end!

You can build a blowgun from a plastic straw in 15 minutes. Spend an afternoon crafting a PVC crossbow. Transform a BBQ lighter into a mini BB shooter in a single sitting!

Ingenuity, action, and explosions await!

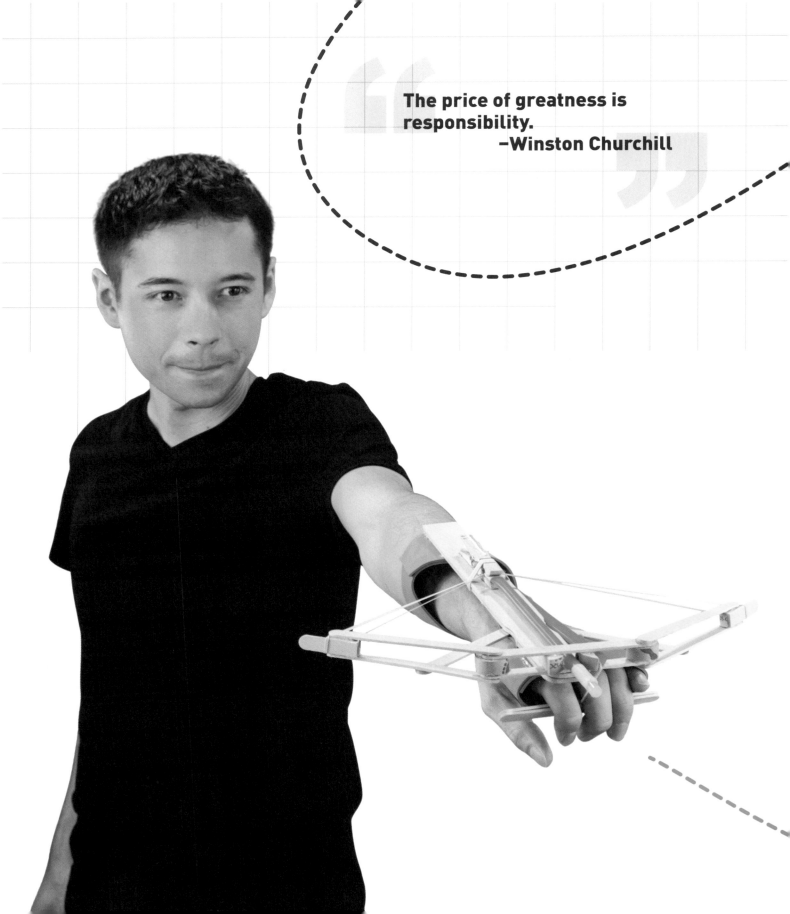

> "The price of greatness is responsibility.
> —Winston Churchill"

GOOD, SAFE FUN

All the projects in this book are designed for fun. Most of them are mildly dangerous. A few of them are extremely dangerous. For me, that's a good thing: Creating crossbows and blasters that have real power inspires the maker in me. It's empowering to know that I built something that can fire more than 100 feet (30 m)! Building something that requires caution and diligence teaches responsibility, a fundamental lesson that all inventors and engineers need to embrace.

The projects with a warning symbol are designed for adults. If you're not quite an adult, I advise creating and using these projects only with adult supervision. If you're the adult supervisor, please pay attention *every* time you fire off one of the high-powered devices. Always use the same amount of caution as you did the first time.

The projects in this book that are deemed Extremely Dangerous have been marked with a WARNING SYMBOL and some information on how to stay safe while operating them.

Please read each project's directions and warnings carefully. Be sure to launch, lob, shoot, and blast off in an area where no one and nothing can be hurt. Be safe—and have fun!

Disclaimer
The reader assumes all responsibility and risk for the use of the advice, information, and directions in or through this book. The Quarto Group; the author, Lance Akiyama; and any affiliated parties do not assume any liability for the content provided in, or available through, this book.

1
SIMPLE AND SUCCESSFUL

STRAW BLOWGUN

Start your ballistic warm-up with this quick and surprisingly gratifying blowgun! The steps are simple, but take your time crafting the darts with care to get the most accurate shots. When you're finished, you'll enjoy the satisfaction of hearing the *thwack!* of a dart piercing cardboard from 10' (3 m) away!

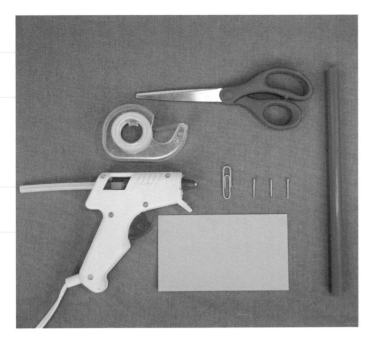

Tools and Materials

- large milkshake straw
- 3 very small ¾" (≈ 2 cm) nails
- paper clip
- index card or similarly sized stiff piece of paper
- scissors
- clear tape or masking tape
- hot glue
- cardboard target or box

MATERIAL SUBSTITUTIONS

Large milkshake straw: You can use any tube that has an inner diameter of about ½" (1.3 cm) and a length of at least 6" (15 cm). Depending on what the tube is made of, you may need a power drill to drill holes if you can't poke holes through it with a nail.

Very small ¾" (≈ 2 cm) nail: You can use other small finishing nails that range from 1" to 2" (2.5 to 5 cm), but be sure to use wire cutters to trim the length down to about ¾" (2 cm). (Always wear protective eyewear when snipping small items.) If the nail is too heavy or long, the dart won't work as well.

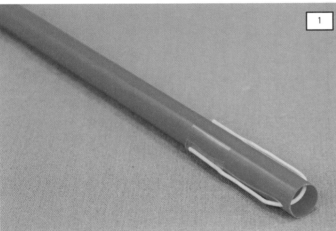

1 Make the blow tube.
Use a nail to pierce all the way through one end of the straw. Straighten the paper clip and thread it through the two holes. Center it and bend both ends along the length of the straw, as shown. Tape the ends in place. This will prevent you from accidentally inhaling a dart!

2 Make the darts.
With the scissors, cut a 1½" × 1½" (4 × 4 cm) square from the index card. Cut a slit from one corner of the square to the center.

SIMPLE AND SUCCESSFUL

3. Cut a small piece of tape and set it aside. Form the paper square into a cone shape by overlapping the sides of the slit, as shown.

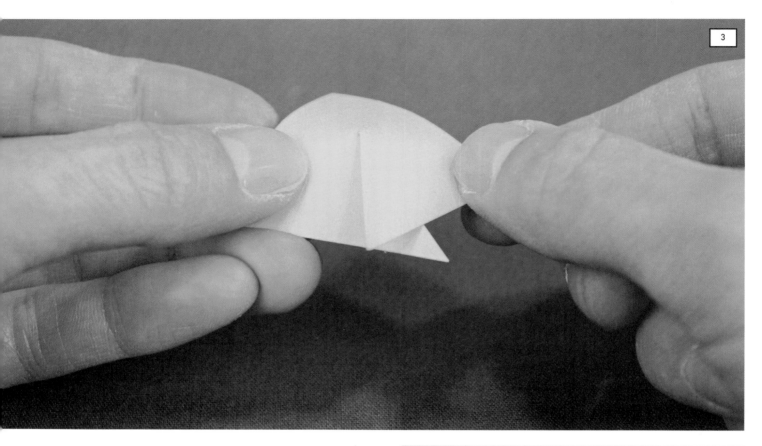

4. Continue wrapping the two sides together, tighter and tighter, until a pointed cone shape is formed. Use one hand to hold the cone in shape and the other to apply the prepared piece of tape. Wrap the cone tightly with tape.

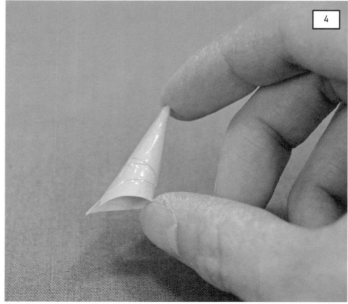

5

Trim the bottom of the cone so the diameter is slightly smaller than the middle of the blow tube. Cut off the very tip of the cone and push the nail through the hole. Fit the nozzle of the glue gun inside the cone and apply a small amount of glue to keep the nail in place. Don't fill the entire cone with glue or the weight will throw off the dart's balance.

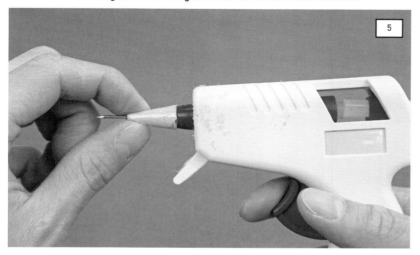

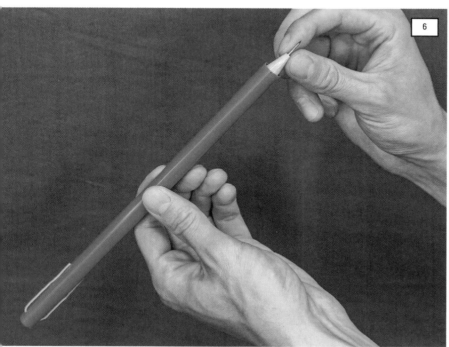

6

Get ready to fire!

Set up a cardboard target on a wall or table and stand 5 to 10 feet (1.5 to 3 m) away from it. Load a dart, tail first, into the blow tube. If it doesn't slide down, gently blow into the open end of the tube to push the dart toward the end with the paper clip. Take aim, and forcefully blow into the tube to send the dart flying!

SIMPLE AND SUCCESSFUL

MICROPULT

You know that sinking feeling when you need a catapult in a hurry and there's none to be found? What to do? Build a micropult! It's quick and simple, yet it packs a punch with a firing distance of more than 15' (4.5 m)! What sets this little lobber apart from other tiny shooters is its basket, which adjusts to neatly fit any small projectile. I used a wad of masking tape for my projectile, but a jelly bean or mini marshmallow would work just as well.

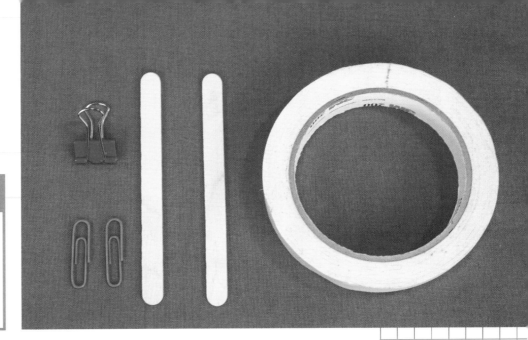

Tools and Materials

- 2 regular 4½" (11.4 cm) craft sticks
- ¾" (2 cm)-wide binder clip
- masking tape
- 2 small paper clips

1

Attach a craft stick to each of the binder clip handles with tightly wrapped masking tape. The upper stick will be the throwing arm and the lower one, the base. Unbend a paper clip, then reshape it into a rectangle about 1" (2.5 cm) long and wide enough to fit over the craft sticks. Position it as shown and tape it to the underside of the base. This will stop the throwing arm from flipping all the way forward and hitting the tabletop. Because the throwing arm stops at a consistent spot, this also makes the micropult more consistent.

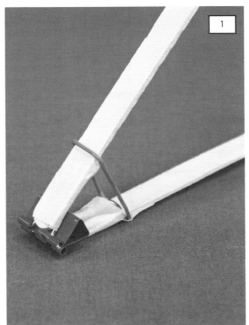

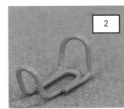

2

Make the projectile basket.
Bend the ends of the second paper clip upward, as shown. Tape the basket onto the end of the throwing arm, about ¼" (6 mm) from the end. You'll need that ¼" to trigger the projectile.

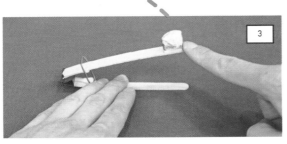

3

You're ready to fire!
Adjust the ends of the basket to snugly (but not tightly) hold your projectile. While stabilizing the micropult with one hand, use a fingertip from the other hand to press down on the end of the throwing arm. Slip your finger off the stick to release!

SIMPLE AND SUCCESSFUL

RUBBER BAND ROCKETS

Transform any pen-like object into a high-flying rocket with nothing but common office supplies. Experiment with different rocket fuselages to find the best performers, then fire up to 50' (15 m) away!

WARNING: Though seemingly simple and harmless, these rockets are fast. Aim carefully!

TOOLS AND MATERIALS

- pen, pencil, craft stick, or another long, slim object
- 1½" (4 cm) brass paper fastener
- masking tape
- index card or card stock
- scissors
- 3½" × ⅛" (9 cm × 3 mm) rubber band

Make the fuselage.

Identify the heavier end of the pencil or pen. This will be the front of the rocket. In my example, the end of the pencil with the eraser is heavier. Attach the paper fastener to the end, as shown, by tightly wrapping it with two pieces of masking tape. Bend the head of the fastener downward slightly; this will make it easier to hook onto the rubber band.

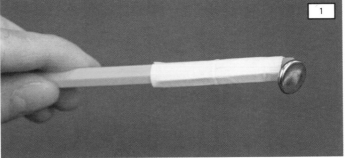

Pro Tip: Use Leading Weights

Rocket-shaped projectiles benefit from having the heaviest part at the front. Imagine trying to throw a plastic straw; it won't go far. Now imagine taping a small rock to one end. When thrown again, the weight of the rock will pull the straw through the air and overcome drag.

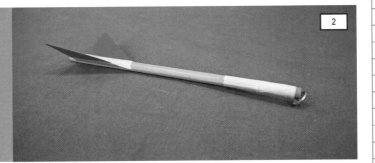

Make the fins.

Cut two triangles from the index card or card stock approximately 1" × 1½" × 2" (2.5 × 4 × 5 cm). Neatly attach the fins near the back of the rocket with masking tape, leaving ½" (1.3 cm) of the pencil exposed; this is where you'll pinch-grip the rocket when drawing it back. Bend the fins up slightly to decrease the chance of the rocket colliding with the rubber band.

SIMPLE AND SUCCESSFUL

Stand by for take off.
Stretch the ends of a rubber band as far apart as you can between your thumb and index finger. While pinching the back of the rocket, hook the fastener head onto the rubber band. Draw it back as far as you can without allowing your fingers to contract toward each other. Take aim and release!

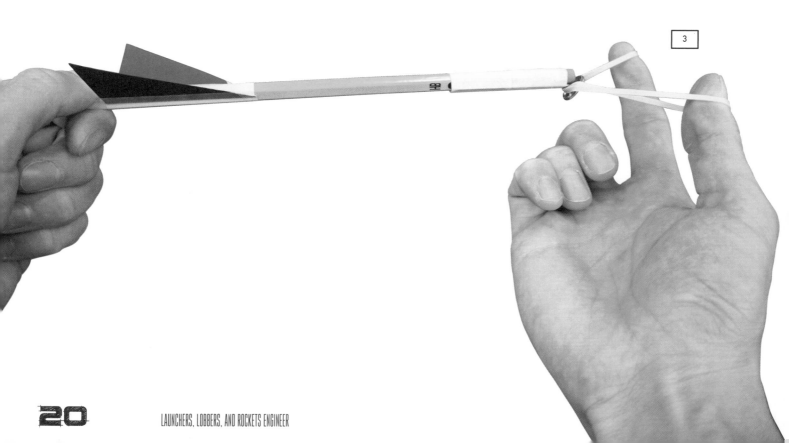

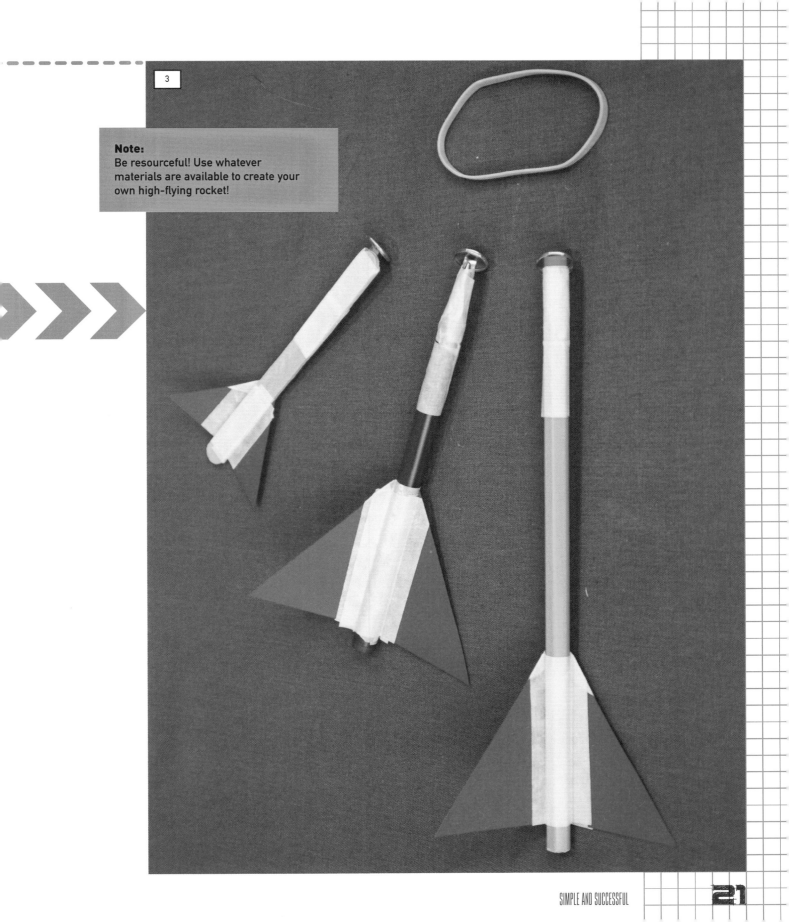

POCKET BOW

Fire straw arrows at whip-cracking speed with enough force to pierce cardboard! The surprising accuracy and consistency of the Pocket Bow is due to the notch in the back of the arrow and the leading weight at the tip. Keep this one in your back pocket for the moment when the need to strike a bull's-eye arises!

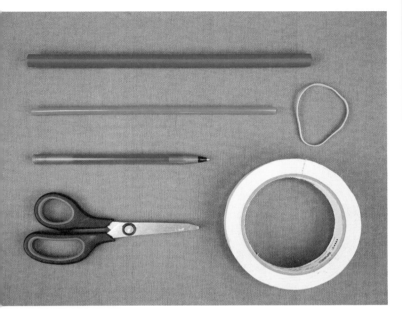

Tools and Materials

- large straw
- masking tape
- scissors
- 3½" × ⅛" (9 cm × 3 mm) rubber band
- smaller straw
- cheap ballpoint pen

MATERIAL SUBSTITUTION
Large milkshake straw: You can use any tube-shaped material that's a little larger than the smaller straw on the list.

1 Make the bow.
Wrap one end of the larger straw with masking tape. Pinch it and cut off the corners to form a point. This notch will help hold the rubber band in position when you're loading the arrow.

SIMPLE AND SUCCESSFUL

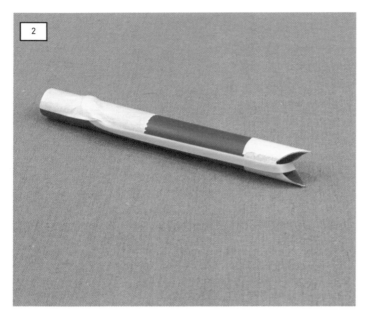

2
Tape a rubber band near the end of the straw so that when it's stretched, it fits neatly into the notch. Cut off the excess straw to shorten the "bow."

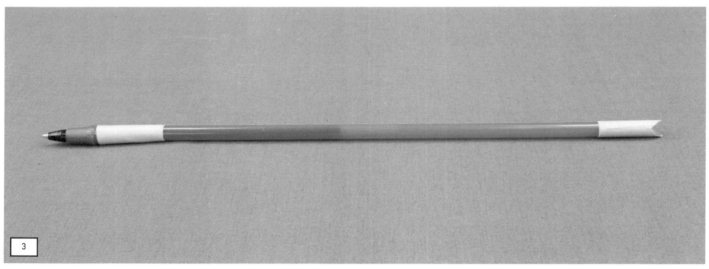

3 Make an arrow.
Create a notch in one end of the smaller straw using the same technique as in step 1. Remove the ballpoint and ink tube from the pen casing. Insert it into the straw, leaving the ballpoint tip exposed. Attach it to the straw with tightly wrapped tape. As with the Rubber Band Rockets project (page 18), having a leading weight increases speed and accuracy.

4

To fire, simply insert the arrow through the front of the bow and hook the nock onto the rubber band. Pinch the back of the arrow in one hand, and lightly grasp the bow in the other. Pull back as far as you'd like—and fire!

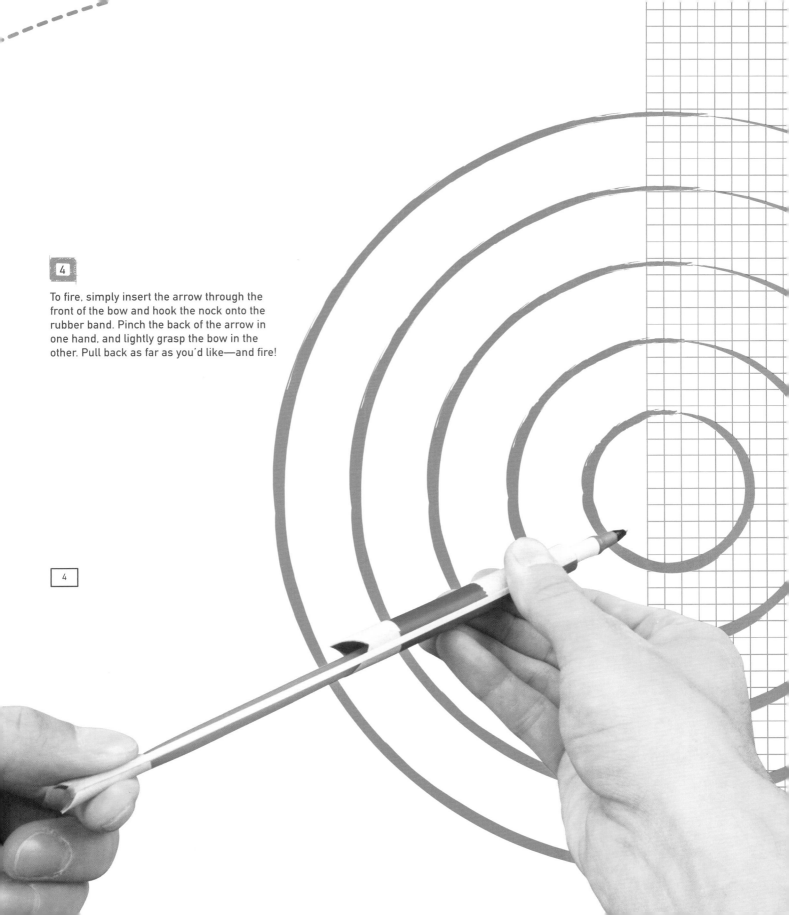

2
BOWS AND SLINGSHOTS

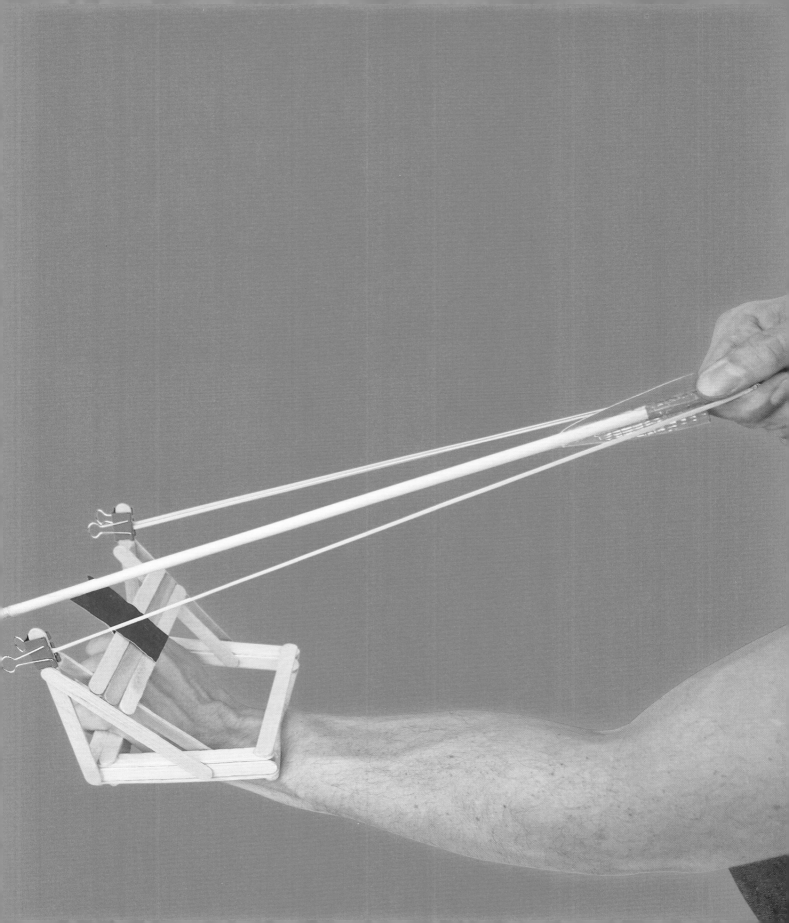

PULLEY-POWERED PVC BOW & ARROW

Drawing inspiration from compound bows, this design uses a simple pulley system to create a mechanical advantage. It makes it easier to draw the bow, while also storing energy efficiently because the bow limbs are pulled directly toward each other. (It also looks way cooler than a regular bow.) When tightly strung, you can fire an arrow 15' (4.6 m) in a straight and speedy line, or aim up and let loose an arc that can reach more than 50' (15 m).

WARNING: This project is exceptionally dangerous! This bow has a 10 to 15 pound (4.5 to 6.8 kg) draw weight, which is enough force to fire the arrows a half-inch (1.3-cm) deep into a plank of wood. Always load and aim while pointing away from people. Never fire directly at a hard surface that may cause your shot to ricochet. You can increase the safety of this project by not sharpening the tip of the arrow.

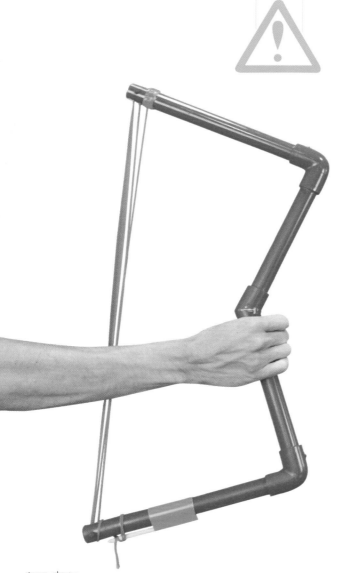

Tools and Materials

- 32" (81.3 cm) of ½" (1.3 cm) schedule 40 PVC pipe
- ruler
- hacksaw or other tool for cutting PVC and wooden dowels
- PVC primer and glue
- ½" (1.3 cm) PVC 45° elbow
- two ½" (1.3 cm) PVC 90° elbows
- needle-nose pliers
- drill with ¼" (6.35 mm) bit
- 24" (61 cm)-long, ¼" (6.35 mm)-thick dowel
- duct tape
- hot glue
- nylon mason line, size #18
- drinking straw that's larger than ¼" (6.35 mm)
- scissors
- 22-gauge floral wire
- index card or card stock
- permanent marker
- spray paint (optional)

MATERIAL SUBSTITUTIONS

Pencil sharpener: You can whittle or saw the dowels to a point.

Floral wire: The wire adds weight to the tip of the arrow. You can substitute by simply adding layers of duct tape. You can also try adhering something else that adds a small amount of weight, such as a small handful of paper clips taped to the arrow.

Nylon mason line: You can use regular cotton string or another nonstretch string, however, size #18 nylon line is the best because of its incredibly high tensile strength, resistance to fraying, and low friction.

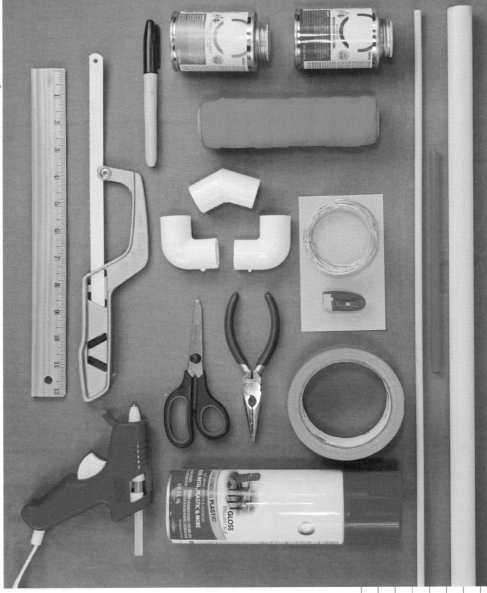

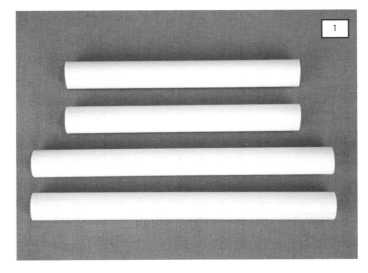

1 Create the bow.
Cut the PVC into two 7" (18 cm) and two 9" (23 cm) pieces. Clear away the debris.

Note:
For the next step, follow the instructions on your PVC primer and adhesive. Cover your work surface with disposable paper. Always wear gloves, and work in a well-ventilated area.

2

Apply primer to the 45-degree elbow and one of the 7" (17.8 cm) pieces of pipe as shown. Work quickly: The adhesive typically sets in about 30 seconds.

To avoid a skewed bow limb: As you assemble the glued pipes, press the bow against a hard, flat surface to ensure that all the pipe pieces are aligned.

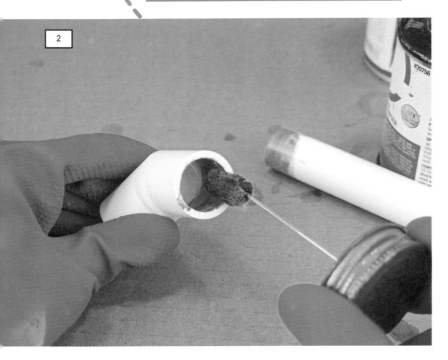

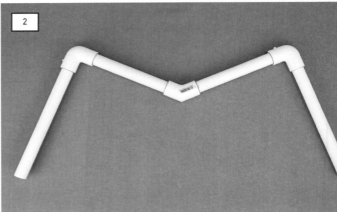

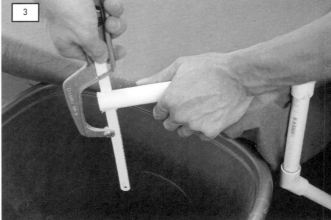

3

Measure and mark a ¼" (6.35 mm)-wide and ¾" (1.9 cm)-deep gap at the end of each limb. Use one hand to hold the bow while it hangs over the corner of your work table. The weight of the bow hanging off the edge will help keep it perfectly vertical. Saw through the pipe.

4 Use needle-nose pliers to tear off any leftover PVC in the notch. Clear away the debris.

5 Drill a hole through the gap, about ¼" (6.35 mm) from the ends of the limbs as shown. Cut two 1½" (4 cm) pieces of dowel and insert them through the holes.

6 (Optional) Spray paint the bow to give it a polished finish. Avoid spraying paint on the parts of the dowels that are inside the pipe.

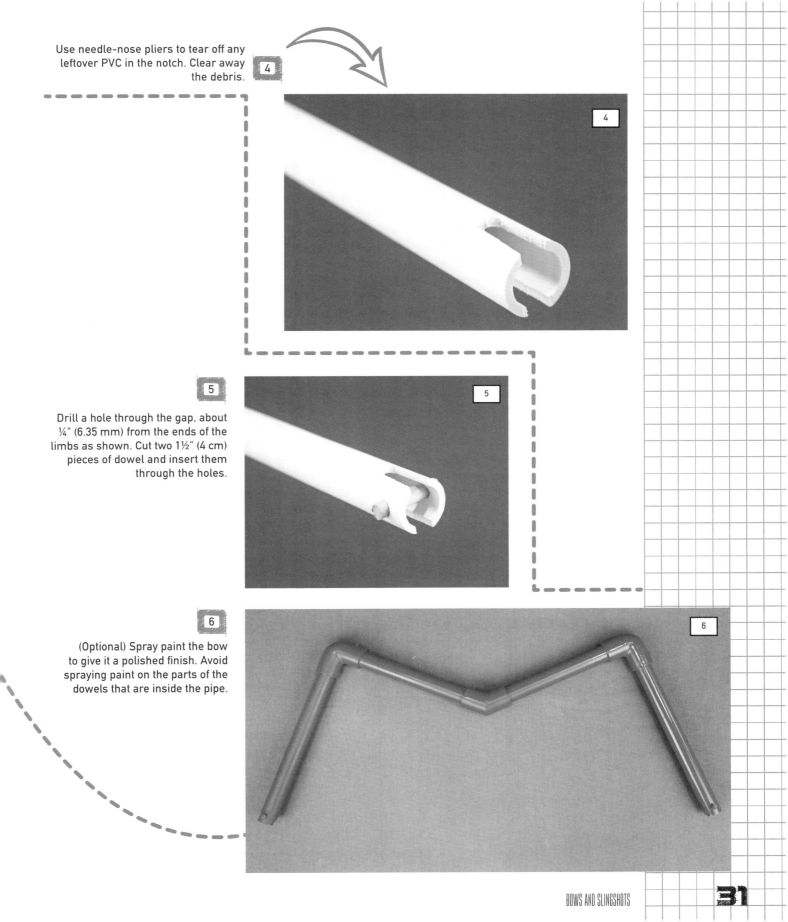

BOWS AND SLINGSHOTS

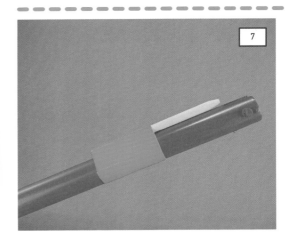

Note:
Read the following steps carefully to ensure that the bow is strung correctly, otherwise you might end up with a pulley system that creates a lot of friction.

7

String the bow.
Cut a dowel to 4" (10 cm), and sharpen it to a blunt point with the pencil sharpener. Hot glue the **nontapered** half of the dowel to the outside of one of the bow limbs (it doesn't matter which one). Don't glue the whole dowel, just the half that's covered in tape. The tip of the dowel should be about ¾" (1.9 cm) from the end of the bow limb. Further secure it with a piece of duct tape.

8 On the other limb, glue the end of the nylon line. Wrap the line around the limb at least five times. Apply more hot glue over the wrapped line to prevent it from unwrapping. Note where the string exits on the side of the limb. As shown here, the string is exiting on the left side, facing you.

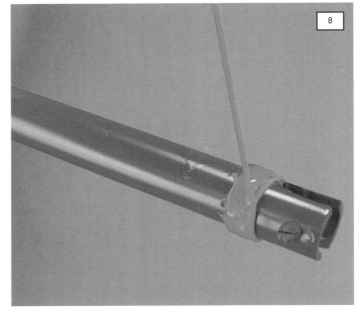

9 Pull the line to the opposite limb, and wrap it around the dowel. The line goes through the back of the dowel, then over the top, as shown.

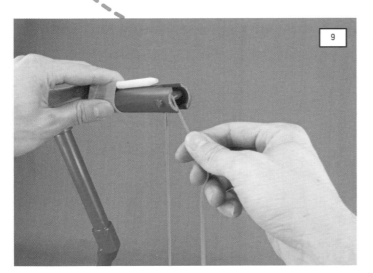

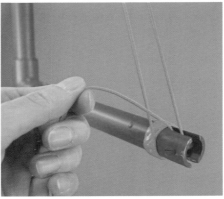

10 Pull the line back toward where it started. Wrap it around this dowel as well, but this time, wrap it from the front first, as shown.

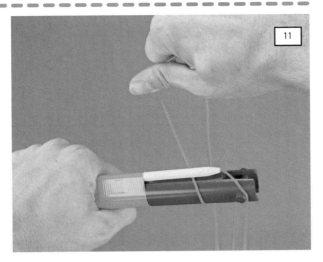

11

Add tension and finish stringing the bow. Use one hand to press down on the bow, bending the limbs together slightly. Adding tension to the bow before finishing the stringing process will generate more power.

With your other hand, pull the line tightly around the outside end of the short 1½" (4 cm) dowel, as shown. It's important to wrap the line around the same side of the bow that was noted in step 8. In this case, it's the left side, facing you. If the line is wrapped from the other side, it will get in the way of the arrow.

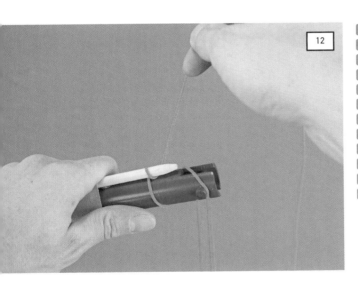

12

Wrap the line around the limb once and slip it under the 4" (10 cm) tapered dowel. The tapered dowel end should make this step easy.

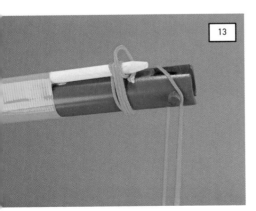

13

Wrap the line around the dowel and limb at least another three times, then trim it. Apply a dab of hot glue on the end of the line to prevent it from fraying, then tuck it snugly under the dowel.

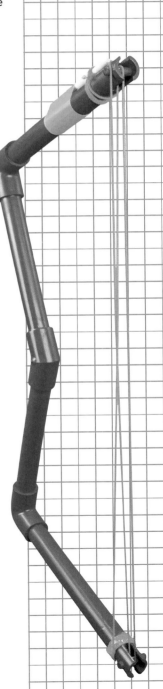

The stringing process is finished! The line is held in place using only friction, yet it's very reliable. Because it's not fixed in place, you can easily unstring the bow, and even adjust the power by changing how much the limbs are compressed before finishing the stringing process.

The stringing process is somewhat precise, so reference this photo as you go.

1 **Make the Arrow**
Cut a 2" (5 cm) piece of straw, and insert an 18" (46 cm) piece of dowel about halfway into it. Wrap it in duct tape.

2 Pinch the end of the piece of straw and cut off the corners to create the arrow nock.

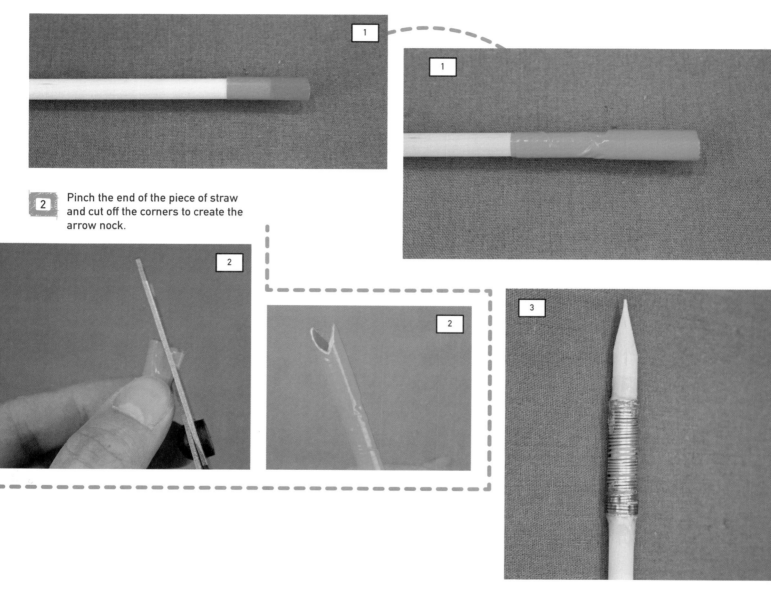

3 Sharpen the other end with the pencil sharpener. Wrap about 1¼" (3 cm) of the dowel with wire. Apply hot glue to prevent the wire from slipping off.

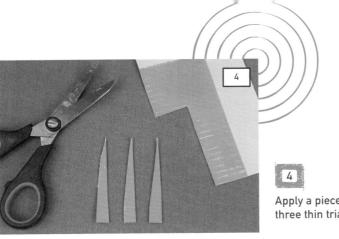

4

Apply a piece of duct tape to both sides of an index card. Cut out three thin triangles, about ¼" × 2½" (6 mm × 6.4 cm).

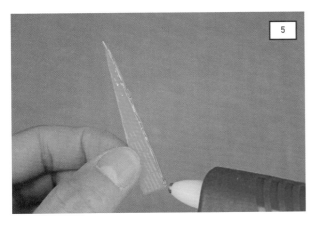

5

Carefully apply hot glue along the straight edge of the fletching. Adhere all three to the back of the arrow, about 1" (2.5 cm) from the nock. Space them out evenly.

Note:
Arrows always have an index fletching. The index fletching is always at a right angle to the slot in the nock. In this picture, it's marked with a black line. Read on to the **Load and fire!** step to better understand the importance of the index fletching.

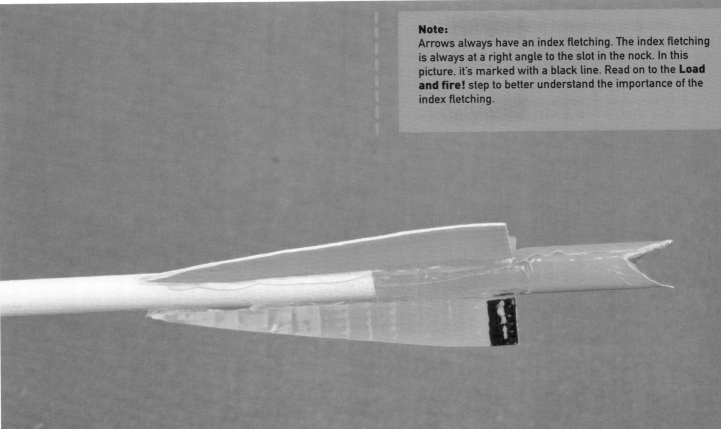

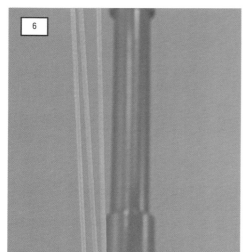

Identify the firing string.
Examine the three bowstrings: They will cross once, but otherwise should not interfere with each other. You'll notice that only one of them is wrapped around the 1½" (4 cm) dowels on both sides. The other two are affixed to one of the limbs. The string that touches both dowels is the firing string. When looking at the strings, as in this photo, the firing string should either be on the far left or far right, never in the middle.

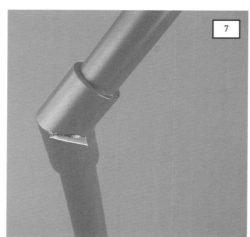

Add the arrow rest.
This is just a very small scrap of taped index card that's glued to the 45-degree elbow. Make sure that it's on the same side of the bow as the firing string.

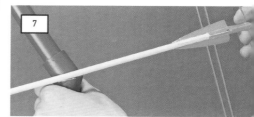

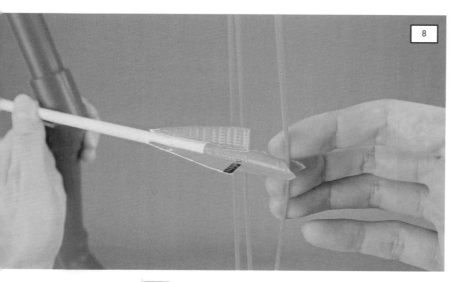

Tip: Left or Right Handed?
The firing string should be on the side opposite your dominant hand. In my example (image 8), the firing string is the rightmost string, and I am right handed. Therefore, I flipped the bow upside down so that the firing string would be on the left. Notice that the 4" (10 cm) tapered dowel is now on the bottom of the bow in the picture on the opposite page. If you're left handed, then the firing string would be the rightmost of the three strings. Make sure you identify the correct bow orientation to match your handedness before adding the arrow rest.

Test the positioning of the index fletching. Nock the arrow on the firing string. The index fletching should be perpendicular (at a 90-degree angle) to the string. The other two fletchings rub lightly against the other strings and may also brush against the bow when fired; this is normal.

Positioning the fletchings like this will prevent them from striking the bow or bowstring when fired. This will ensure that your shot is straight and the fletching doesn't get damaged.

TROUBLESHOOTING

- If the bow is lacking power, be sure to bend the bow slightly while stringing it and that the string is taut throughout the stringing process.

- Double check that you strung the bow correctly. Check for excessive friction or pinch points that are hindering the movement of the string.

- Make sure you're using the firing string. Remember to pick the one that is wrapped around both pulley dowels.

- If the arrow wobbles during flight, try adding more weight to the tip. Also check your fletching: Make sure it's attached perfectly straight and is undamaged.

- Still wobbling? The fletching may be striking the bow or the other strings. Trim your fletching so that it's no more than ¼" (6.35 mm) tall. Double-check that the index fletching is in the right position.

Load and fire! Nock the arrow onto the firing string and place the shaft onto the arrow rest. Hold the nock in place between your index and middle fingers. Hold the firing string with just your fingertips; don't grab or grip the string.

Pull the firing string back until the tip of the arrow is just in front of the hand gripping the bow.

Never draw the arrow so the point is behind your hand, or you may accidentally shoot yourself!

You'll notice that the bow's shape transforms as energy is stored in the bent PVC!

For the most accurate shot, don't pluck or pull at the string when releasing. Instead, release by relaxing your fingers and allowing the bowstring to slip away with a *twang!*

Pro Tip: Take out the Tension
Undo the bowstring once you're done shooting for the day. If the bow is stored for long periods while under tension, the PVC limbs will permanently bend, thus reducing the maximum power you can achieve.

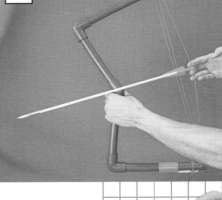

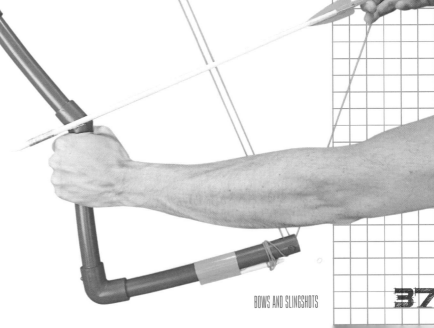

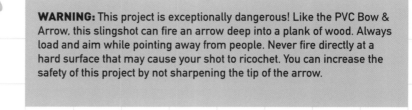

WARNING: This project is exceptionally dangerous! Like the PVC Bow & Arrow, this slingshot can fire an arrow deep into a plank of wood. Always load and aim while pointing away from people. Never fire directly at a hard surface that may cause your shot to ricochet. You can increase the safety of this project by not sharpening the tip of the arrow.

SLINGSHOT AND ARROW

You've heard of the bow and arrow, but what about the slingshot and arrow? This easy-to-build contraption packs all the punch of a bow but in a smaller, handheld-sized package. With just one long rubber band, you'll be accurately firing arrows at least 40' (12 m)!

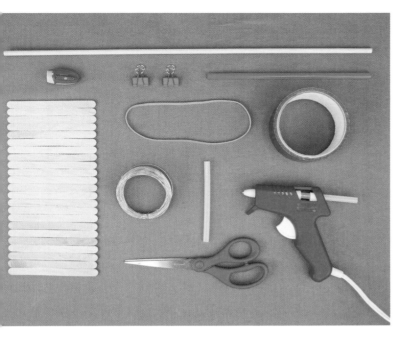

Tools and Materials

- scissors
- mini hot glue stick
- hot glue
- 23 regular 4½" × ⅜" (11.4 × 1 cm) craft sticks
- duct tape
- 7" × ⅛" (18 cm × 3 mm) size #117b rubber band
- two ¾" (1.9 cm) binder clips
- 18" (45.7 cm)-long, ¼" (6.35 mm)-thick dowel
- drinking straw that's wider than ¼" (6.35 mm)
- 22-gauge floral wire
- index card or card stock
- permanent marker

MATERIAL SUBSTITUTIONS
7" × ⅛" (18 cm × 3 mm) size #117b rubber band: You can tie two regular 3½" × ⅛" (9 cm × 3 mm) size #33 rubber bands together.

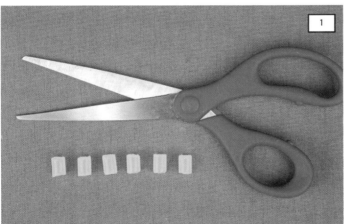

1 **Create the slingshot frame.** Cut off six ⅜" (1 cm) pieces of a mini glue stick.

2 Hot glue four craft sticks together into a square. Glue four of the glue stick pieces onto the corners as shown.

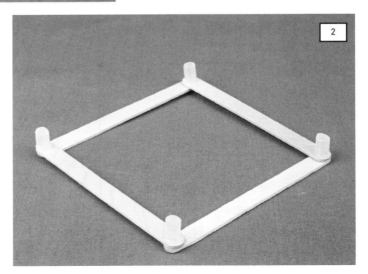

3. Glue more craft sticks onto the remaining three sides of the glue stick pieces.

Glue on additional craft sticks, as shown, to strengthen the frame. The side that has two pairs of craft sticks glued side-by-side will be the handle.

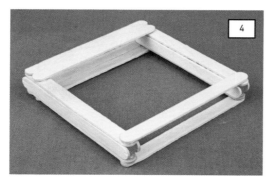

Create the slingshot supports.
Glue two sticks upright at a 90-degree angle next to the handle. Support those with two diagonal sticks as shown. Be sure to leave about 1" (2.5 cm) of the upright craft sticks exposed above the point where the diagonal sticks connect.

6

Glue the remaining two pieces of mini glue stick onto the uprights, opposite the point where the diagonal sticks connect. Attach another pair of diagonal sticks, as shown. This will further reinforce the uprights.

Apply hot glue on the edges of the innermost diagonal sticks and, working quickly, attach three sticks, as shown. These will prevent the uprights from bending inward and will provide a place to attach the arrow rest.

8

Create the arrow rest.
Cut two 3" (7.5 cm) pieces of duct tape and align the sticky sides together. Cut the piece in half to create two pieces approximately 3" × 1" (5 × 2.5 cm). Cut out a shallow triangle on one end of both.

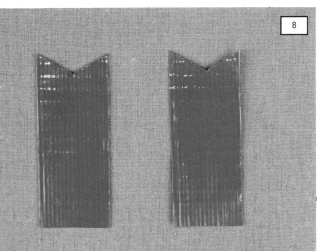

BOWS AND SLINGSHOTS

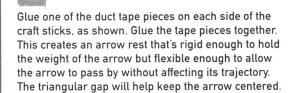

9 Glue one of the duct tape pieces on each side of the craft sticks, as shown. Glue the tape pieces together. This creates an arrow rest that's rigid enough to hold the weight of the arrow but flexible enough to allow the arrow to pass by without affecting its trajectory. The triangular gap will help keep the arrow centered.

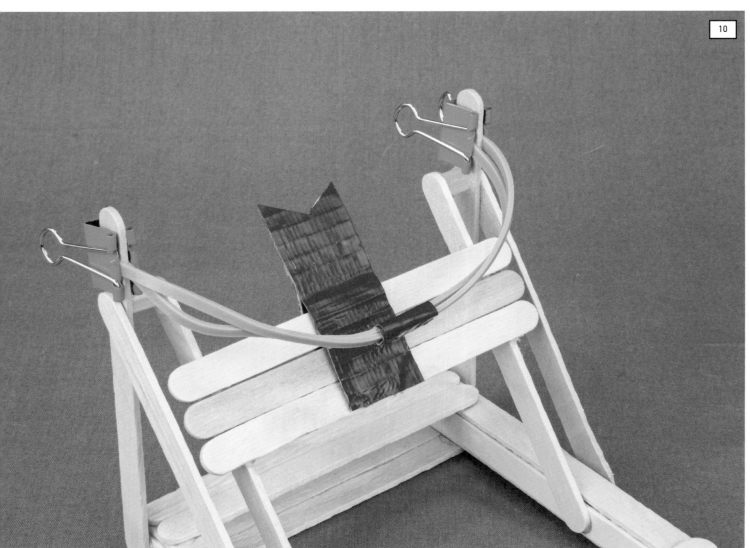

10 Add the firepower! Clip on the 7" × 1/8" (18 cm × 3 mm) rubber band with the two binder clips as shown. Tape the middle of the rubber band together with a small piece of duct tape.

11 Create an arrow. Follow the same steps for making an arrow as shown in the Pulley-Powered PVC Bow & Arrow on pages 34–35.

Load and fire!
Slip your nondominant hand under the slingshot frame, and grasp the handle. Hold the back of the arrow in your dominant hand and fit the nock over the duct taped part of the rubber band.

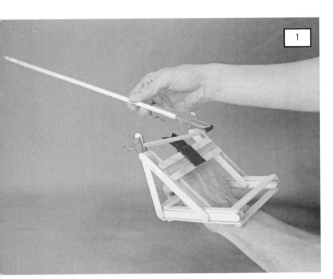

Temporarily hold the nocked arrow with the same hand that's grasping the slingshot by fitting it between your fingers. This frees your dominant hand to switch positions.

Firmly pinch the arrow's nock from behind the rubber band, using your dominant hand. Pull back until just 2" to 3" (5 to 7.5 cm) of the arrow remains in front of the slingshot. Keep the arrow positioned in the center of the arrow rest. Take careful aim, and release!

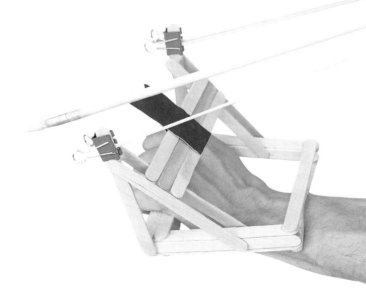

DUCT TAPE & PVC CROSSBOW

There are lots of PVC crossbow how-to tutorials, but I found none that were satisfyingly powerful, simple to build, and constructed primarily out of PVC. This crossbow is all of that *and* stylishly designed! Be warned: This project is relatively powerful and dangerous with a 15-pound (6.8 kg) draw weight. *Fire with safety goggles, and take aim with care!*

WARNING: This project is exceptionally dangerous! A 15-pound (6.8 kg) draw weight has enough force to fire the bolts ½" (1.3 cm) deep into a plank of wood. Additionally, having a trigger mechanism means that there is a chance that the crossbow can be released unintentionally. Always load and aim while pointing away from people. Always double check that the trigger is loaded properly before raising the crossbow to fire. Never fire directly at a hard surface that may cause your shot to ricochet. Never fire at a person or animal. Do not build unless you can operate the crossbow safely.

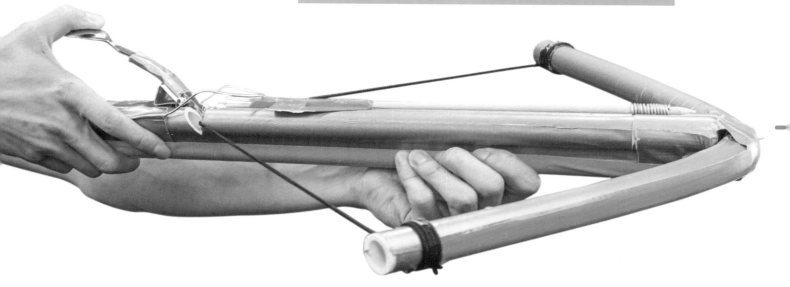

Tools and Materials

- ruler
- PVC pipe cutters or hacksaw
- 84" (2.1 m) of ½" (1.3 cm) schedule 40 PVC pipe*
- heavy-duty duct tape
- 36" (1 m) of paracord
- power drill with a ½" (1.3 cm) bit, and bit that's slightly wider than the paracord
- hot glue gun
- utility knife
- hacksaw or other tool for cutting into PVC (not PVC cutters)
- safety goggles
- 1½" (4 cm)-wide binder clip
- metal spoon**
- 16-gauge mechanic's wire
- 12" (30 cm)-long, ¼" (6 mm)-wide wooden dowel
- pencil sharpener
- index card
- decorative duct tape (optional)

*Use only new PVC pipe. Don't use pipe that has been exposed to the sun or has been used in other applications.

**Use a spoon with a wide, flat handle that is difficult to bend.

 Use the PVC cutters or a hacksaw to cut the PVC into three 20" (51 cm) pieces for the stock, and one 24" (61 cm) piece for the bow.

Assemble the stock.
Bundle the three 20" (51 cm) pieces of PVC, then wrap the ends tightly in duct tape. Press the tape into the grooves as you wrap it around. This will ensure that at least one side will have a grooved channel to help guide the crossbow bolt.

Apply three 20" (51 cm) pieces of tape along all three grooves in the stock. Again, press the tape into the grooves.

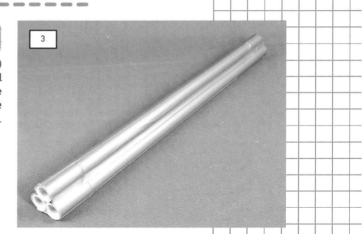

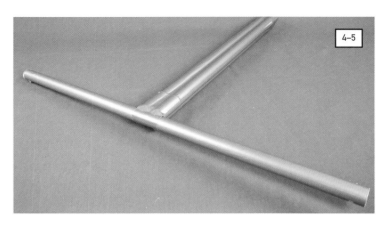

4 Prepare the bow by covering the 24" (61 cm) piece of PVC with duct tape. The tape will help prevent the bow from releasing shrapnel in the unlikely event that it breaks. Plus, it looks way better!

5 **Attach the bow.** Measure and mark the center of the bow. Line it up with one end of the stock. Tightly wrap a piece of duct tape around the bow and the front of the stock, securing the bow in place. Again, press the tape into the groove.

6 Wrap a second piece of tape around the first piece, all the way around the front of the stock and press it into the groove. (I've photographed this step with white duct tape to better show where the tape is being added, but you can use any color you like.)

7 Repeat the taping method from the last two steps to further secure the joint. This part of the crossbow will be under stress, so it's important to layer the tape to prevent it from ripping.

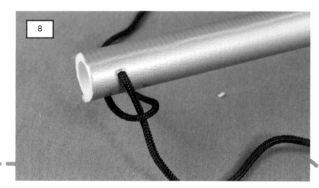

8 **String the bow.** Using a drill with a bit that's slightly wider than your paracord, drill a hole into each end of the bow. Insert one end of the paracord into the hole and push it toward the front of the bow. Use a bit of wire to help push it through because it'll be a tight fit. Loop one end of the paracord around the other, as shown.

9 Make sure the paracord is pulled tightly toward the underside of the bow. This will help ensure the bolt stays in the groove during firing. Wrap the paracord tightly around the bow several times, but do not wrap it around itself. On the last wrap, apply hot glue to the bow and adhere the paracord to it.

10 Allow the glue to dry completely, then repeat on the other side while pulling the paracord taut. It's easiest to do this step with a friend: Have one person keep the paracord taut while the other wraps the paracord and glues it in place. The bowstring is done!

11

Create the trigger notch.
Measure 14" (35.5 cm) from the very front of the bow and mark the spot on top of the stock. Using the utility knife, cut away about ½" (1.27 cm) of duct tape between 14" and 14½" (35.5 and 36.8 cm). This will prevent the tape from gumming up the saw blade.

12

Saw straight down about ½" (1.3 cm) into the stock at the 14" (36 cm) mark. Now saw diagonally at the 14½" (37 cm) mark toward the first cut, as shown. Clean up the PVC debris.

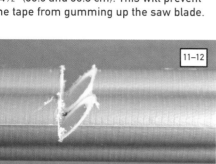

13

Test the trigger notch.
On a solid, nonslip surface, press the end of the stock into the ground and grasp the bowstring with both hands. Slowly push the bowstring down, as shown, until it slips into the trigger notch. It will require upward of 40 pounds (18 kg) of force to string the bow the first time. It might be a little unnerving to see the PVC bend so much: Put on your goggles and push hard!

14

Leave the bowstring in the trigger notch for 5 minutes. This will help the PVC bend into its resting state, making it much easier to load in the future. Release the string by inserting the handle of the metal spoon into the trigger notch and using it as a lever. Lift the string out of the notch with a snap! Expect the bowstring to be a little slack now.

15

Make the trigger.
Cut away a little more tape at the trigger notch, where the trigger hole will be drilled as shown. Using the ½" (1.3 cm) drill bit, drill a hole into the center of the stock in front of the notch. This step may be challenging because the drill bit may slip out of place at first. To prevent this, tightly grasp the stock with one hand to prevent the pipe from being pushed apart by the force of the drill. Work slowly and diligently.

16

Tape the handle of the binder clip onto the stock as shown. Tape a spoon onto the upper handle so that the end of the spoon's handle rests just below the trigger hole.

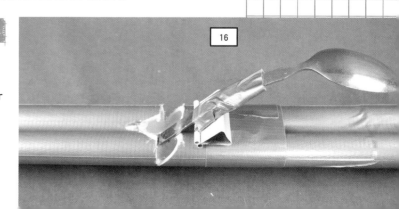

17. Push on the spoon to test that the handle lifts through the trigger hole. If desired, bend the spoon into a more comfortable position, as shown.

Build the bolt clip.
18. Cut an 8" (20 cm) piece of wire and tape the ends near the sides of the binder clip. Bend the wire so it's pressed into the center of the stock. This will keep bolts in place while you take aim. The crossbow is done!

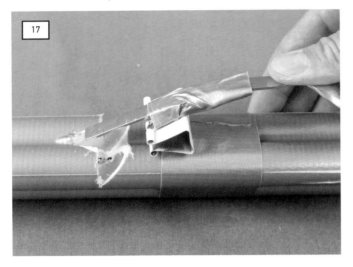

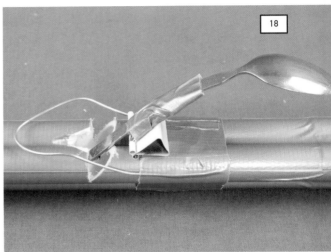

Make the Crossbow Bolts
1. Split one end of the dowel with a utility knife. Carefully insert a small piece of wood or tiny pebble, as shown.

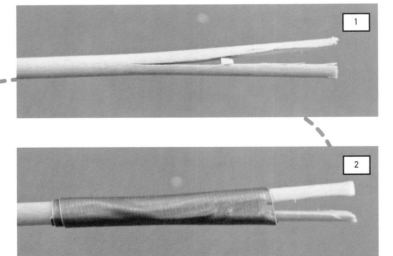

2. Wrap duct tape tightly around the split, leaving about 1" (2.5 cm) of space for the nock. The nock is needed to catch onto the bowstring.

3 Sharpen one end of the dowel with a pencil sharpener. Wrap about 5" (12.5 cm) of wire around the tip of the bolt. The added leading weight will help pull the bolt through the air with its momentum.

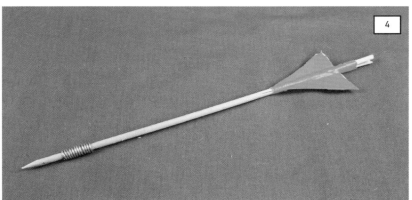

4 Cut two triangular fins from an index card, each about 2½" (6.4 cm) tall and 1½" (4 cm) wide. Tape them to the end of the bolt. Make sure the fins are aligned with the nock, as shown. Be sure they are directly opposite each other and perfectly straight. Cover the fins with duct tape for added durability.

Pro Tip: Safer Bolt Alternative
Instead of sharpening the tip of the dowel, cover it with a pencil-cap eraser.

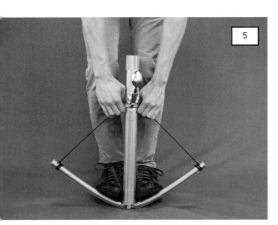

Load and fire.
5 Put the front of the crossbow on the ground, and firmly step on the bow with the tips of both feet, as shown. Using both hands, draw back the bowstring into the trigger notch.

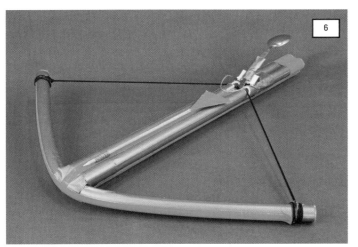

6 Position the bolt under the bolt clip. Make sure the bolt's nock is directly in front of the trigger notch. Keeping your hands clear of the top of the stock, squeeze the spoon trigger to release the bowstring and *fire!*

BOWS AND SLINGSHOTS

WRIST-MOUNTED CROSSBOW

The *twang!* of a crossbow string as it releases is deeply satisfying, and if the crossbow is attached to your arm when you fire it, the *twang* is positively cyborgian! This Wrist-Mounted Crossbow packs plenty of satisfaction and will let you fire bolts up to 40' (12 m) away!

It might take a little extra diligence to get the trigger working just right, but let me assure you: It's worth it!

LAUNCHERS, LOBBERS, AND ROCKETS ENGINEER

Tools and Materials

- two 12" (30 cm) wooden paint stirrers
- hot glue
- eleven regular 4½" (11.4 cm) craft sticks
- utility knife
- wine cork (synthetic preferred)
- two 3½" × ⅛" (9 cm × 3 mm) size #33 rubber bands
- wooden clothespin
- cotton string
- drill with ⅛" (3 mm) bit
- two 8" (20 cm) large straws
- one 8½" × 11" (21.6 × 28 cm) sheet of craft foam
- masking tape
- scissors

MATERIAL SUBSTITUTIONS
Wine cork: The cork is used to create space between layers of craft sticks. You can substitute it with small ½" (1.3 cm) blocks of wood (such as those sawn from a square wooden dowel), or cut and layer bits of craft sticks as shown in the Ballista project on page 74.

Craft foam: Use felt or other soft fabric. You can also use Velcro straps or any other means to strap the crossbow to your wrist.

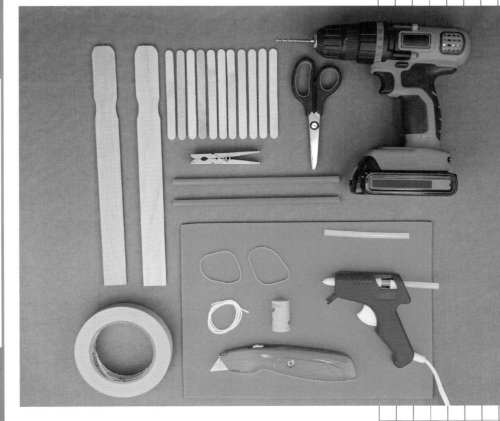

Create the crossbow.

1. If your paint stirrers are thin like mine (about ⅛" [3 mm] thick), then you'll need to hot glue two together to prevent them from bending.

2. Glue three sticks to one end of the paint stirrers, as shown. Place the two outer sticks at a wide angle, about 145 degrees.

3. Use the utility knife to carefully slice the cork into fourths, then cut each of those segments in half.

Note:
The distance between the two center pieces must be at least ⅛" (3 mm) greater than the width of the straw you'll shoot. If the gap is too narrow, the straw will get stuck. In this example, the gap is about ⅜" (1 cm) wide.

4 Glue the cork segments to the craft sticks, as shown.

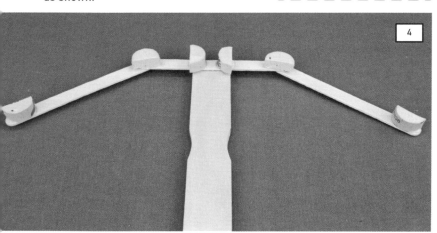

5 Glue another layer of craft sticks in place in the same position as the sticks in step 2. (Two layers of sticks separated by the pieces of cork creates a rigid structure that won't bend easily under the strain of the rubber band.)

Glue two more sticks from the bow to the paint stirrers. These will further prevent the bow from bending under the force of the stretched rubber band.

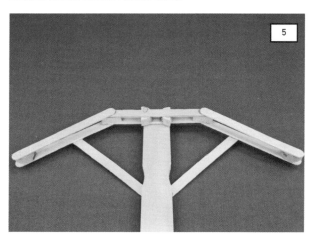

6 String the crossbow.
Cut two small pieces of craft stick, about 1½" (4 cm) long. Glue them onto the flat sides of the outermost cork pieces, as shown. When the glue has fully set, loop the rubber band around one side, give it a single twist, then stretch and loop it onto the other side. Twisting the rubber band will make it easier to load the crossbow later.

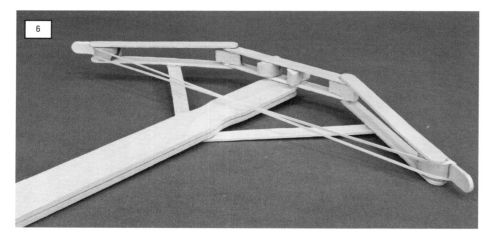

7 Create the trigger.
The wood surface inside a clothespin's pinching end doesn't have much grip. Intensify the grip by applying a small bead of hot glue inside the opening as shown. While the glue is hot, repeatedly open and close the clothespin until the glue dries, taking care not to glue the trigger shut. This will create a textured, rubbery surface that will improve the clothespin's ability to hold onto the bolts.

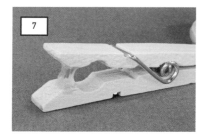

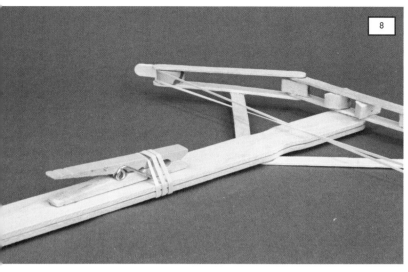

8 Glue the clothespin/trigger onto the paint stirrers. The pinching end of the clothespin should be 6" (15 cm) from the front of the crossbow. Further improve the clothespin's grip by wrapping a rubber band around the opening, as shown.

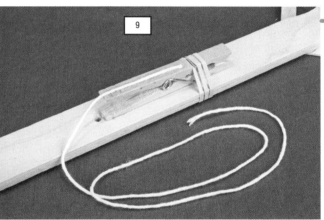

9 Cut a 15" (38 cm) length of string. Glue about 2" (5 cm) of one end of the string to the top of the clothespin. Drill a 1/8" (3 mm) hole through the paint stirrers, directly behind the clothespin.

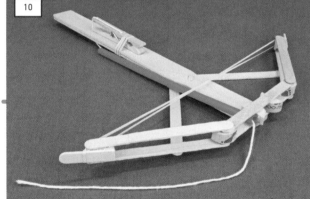

10 Glue one of the straws onto the side of the crossbow, as shown. Make sure the end of the straw is lined up with the front of the crossbow. Thread the string through the hole in the paint stirrers, then through the straw. This will reduce the string's friction, and make it easier to pull the trigger.

Important Note:
When the string is pulled taut, there must be about 1" (2.5 cm) of string between the front of the crossbow and the trigger handle. If the string is too short, you won't be able to fit your fingers over the handle. If it's too long, it will take more effort to open the trigger. Wrap or unwrap string around the craft stick until you have the correct distance.

When you have the right amount of string wrapped around the handle, glue another craft stick onto the first. This will lock the string into place and prevent it from unwinding.

11 Create the trigger handle.
Wrap the remaining 3" (7.5 cm), or so, of string around a craft stick. Apply hot glue and a second craft stick to keep the string in place.

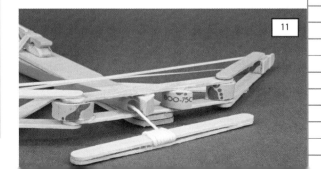

12

Create the straps and padding.
Cut three pieces of craft foam: one 11" × 2½" (28 × 6.4 cm), one 11" × 1½" (28 × 4 cm), and one 1" × 3" (2.5 × 7.5 cm).

13

Wrap the 11" × 1½" (28 × 4 cm) piece of foam around your hand at midpalm. Holding it in position, slip it off the palm, then using hot glue, attach the ends to permanently maintain the shape.

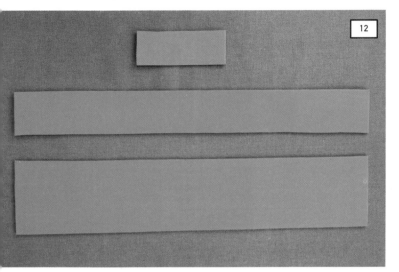

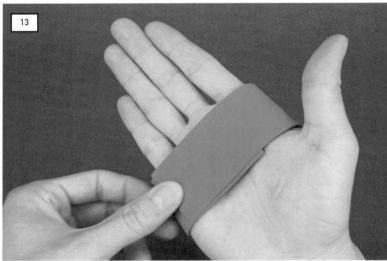

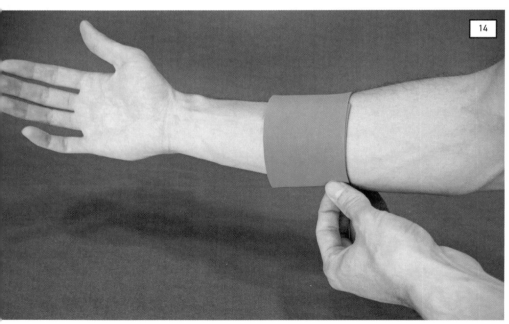

14 Repeat this process with another strap that's about halfway up your forearm using the 11" × 2½" (28 × 6.4 cm) foam strip.

15 Flip the crossbow upside down. Glue the forearm strap at the very back of the crossbow and the palm strap about 1" (2.5 cm) from the front. Glue the 1" × 3" (2.5 × 7.5 cm) piece next to the palm strap to protect your knuckles. Measurements may vary, depending on the shape of your arm.

1 Make the crossbow bolts.
Cut a 1" (2.5 cm) piece of hot glue stick and tape it to one side of the remaining straw. Wrap a layer of masking tape around the other end.

2 Pinch the taped end of the straw and cut off the two corners with scissors. This creates a nock that will fit over the rubber band. The layer of tape will help prevent the straw from splitting over time.

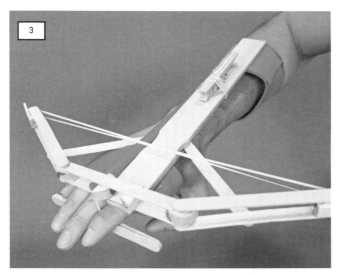

3 Load and fire.
Slip your arm through the straps. Place your fingers over the trigger handle, as shown at left.

TROUBLESHOOTING
The trigger-handle string length and the sensitivity of the clothespin trigger are the two key things you may need to fine tune.

- Make sure the string that connects to the trigger handle is the right length. When your fingers are resting on it, they should be pointing straight forward, as shown in step 3. If not, tighten the string by wrapping it around the handle and gluing it in place. If the string is just 1" (2.5 cm) too long or too short, it won't work well.

- If the clothespin is too difficult to open, try loosening the rubber band that's wrapped around it.

- If the bolt is slipping out from the trigger prematurely, make sure to push the nock as far back into the clothespin as you can.

- Lastly, double-check that the string is not getting caught on hot glue, getting stuck in a crack between two materials, or otherwise encountering excessive friction.

4 Insert a bolt through the front of the crossbow between the cork pieces.

Hold the bolt very close to the nock and fit it onto the rubber band. Open the clothespin by pulling on the trigger handle. Push the bolt toward the open clothespin. Release the trigger handle once the bolt is at least ½" (1.3 cm) inside the trigger.

5 Take careful aim, and give the trigger handle a light squeeze to send your shot flying!

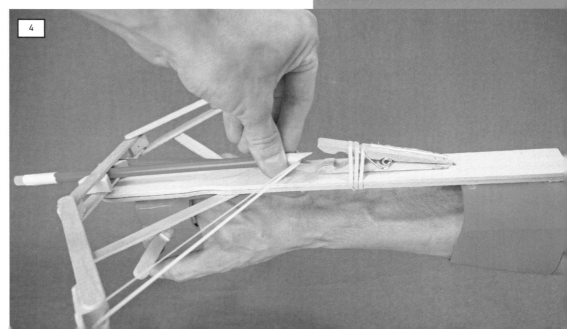

MINI MEDIEVAL SIEGE MACHINES

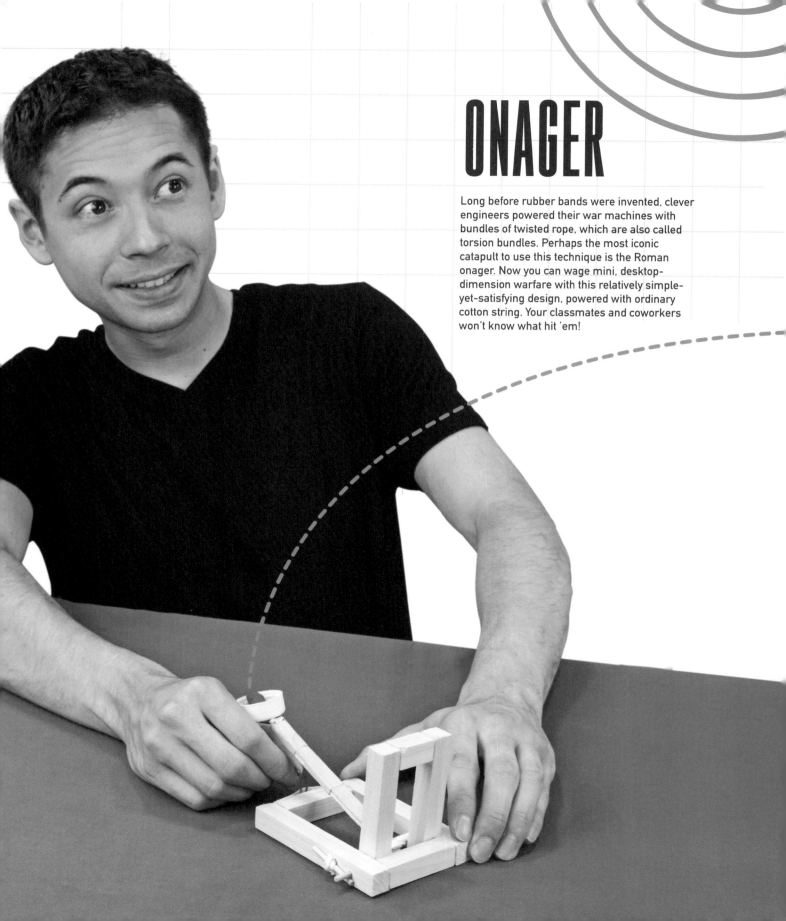

ONAGER

Long before rubber bands were invented, clever engineers powered their war machines with bundles of twisted rope, which are also called torsion bundles. Perhaps the most iconic catapult to use this technique is the Roman onager. Now you can wage mini, desktop-dimension warfare with this relatively simple-yet-satisfying design, powered with ordinary cotton string. Your classmates and coworkers won't know what hit 'em!

Tools and Materials

- ½" (1.27 cm)-thick square wooden dowel, at least 19" (48 cm) long
- saw or another tool to cut the dowel
- hot glue
- drill with ¼" and ⅛" (6 mm and 3 mm) drill bits
- 2 regular ½" × 4½" (1.3 × 11.4 cm) craft sticks
- 4 paper clips
- plastic bottle cap
- masking tape
- cotton string
- ⅛" (3 mm) bamboo skewer
- pencil

MATERIAL NOTES
Bamboo skewers are crucial for this project: Accept no substitutes. The strands of fiber in bamboo are resistant to the crushing force of the torsion bundle. Skewers come in large packs and many are not manufactured well. Take time to select a skewer that seems exceptionally thick and durable.

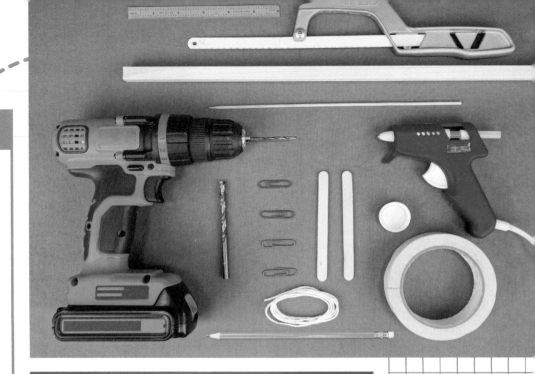

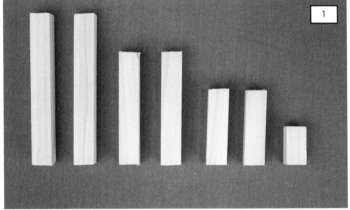

 Cut the square dowel into two 4" (10 cm), two 3" (7.5 cm), two 2" (5 cm), and one 1" (2.5 cm) pieces.

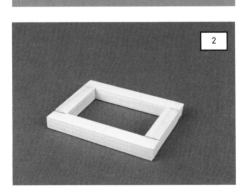

Create the base.
Glue the 4" (10 cm) and 2" (5 cm) pieces of dowel into a rectangle, as shown. Make sure the short pieces are positioned between the long pieces.

Create the upright.
Glue the 1" (2.5 cm) piece of dowel between the two 3" (7.5 cm) pieces, as shown.

MINI MEDIEVAL SIEGE MACHINES

5

Using the ¼" (6 mm) bit, drill holes into the base, directly behind the uprights. Make sure the holes are aligned with each other. Run the drill through each hole several times to remove rough edges and debris as much as possible. The string will feed through these holes.

4

Liberally apply hot glue in two corners of the base. Pinch the bottoms of the upright together slightly and insert it into the base. Push the bottoms of the upright firmly into the glued corners.

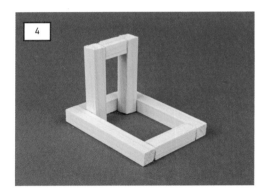

6

Using the ⅛" (3 mm) bit, drill a ½" (1.3 cm)-deep hole into the base, alongside of each upright. These holes will hold pegs for the winding mechanism.

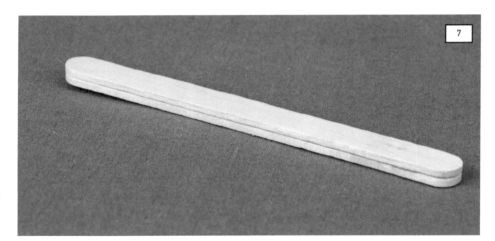

7

Create the throwing arm.
Glue two craft sticks together.

8

Glue and tape two paper clips to one end of the craft sticks.

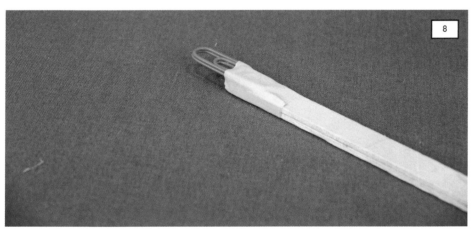

Note:
If only one paper clip is used, the bottle cap will not hold its position each time the catapult is fired.

LAUNCHERS, LOBBERS, AND ROCKETS ENGINEER

9

Glue a plastic bottle cap to the paper clips. Keep adding layers of glue until the paper clips are completely encased in glue.

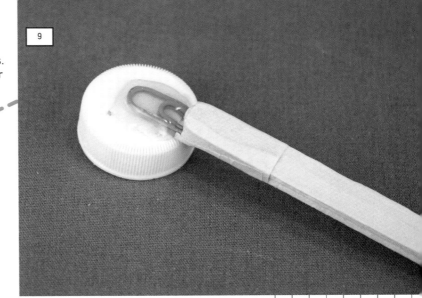

10

Create the torsion spring. Cut a 44" (112 cm) piece of cotton string. Tie the ends together.

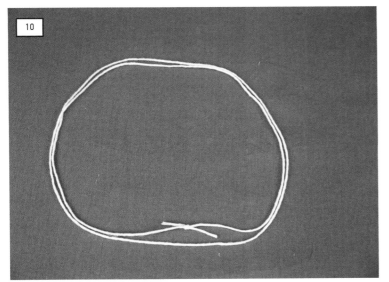

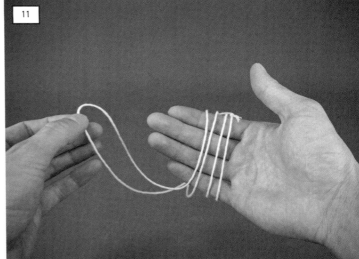

11

Wrap the string around your hand or something else that's about 4" (10 cm) wide. You should be able to wrap it five times. Set the looped string aside.

12

Cut the bamboo skewer into two 1⅜" (3.5 cm) pieces and two 1" (2.5 cm) pieces.

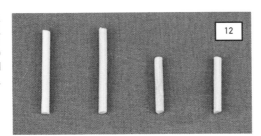

MINI MEDIEVAL SIEGE MACHINES

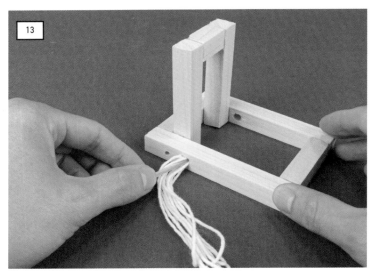

13 Use a piece of skewer to push the looped string through the ¼" (6 mm) holes.

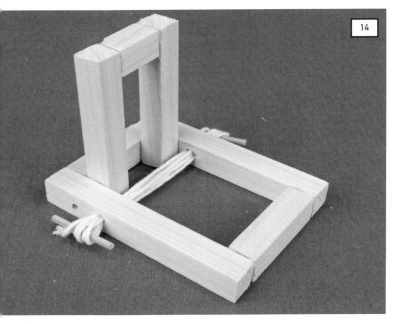

14 Straighten the string as much as you can. Insert a 1⅜" (3.5 cm) skewer through all of the looped ends of the string. These will be the winding pegs. Count to make sure all five strands of string are wrapped around each peg.

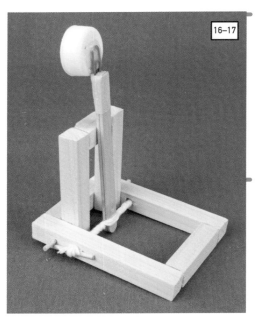

16 Begin turning the winding pegs to tighten the torsion bundle. Tighten each side evenly.

15 Install the throwing arm.
Insert the throwing arm into the middle of the string bundle. Leave about ¼" (6 mm) poking out the other side of the string.

17 Wind until the throwing arm is pressed firmly against the uprights. Insert the 1" (2.5 cm) holding pegs into the ⅛" (3 mm) holes to prevent the winding pegs from unwinding. You'll add more power later.

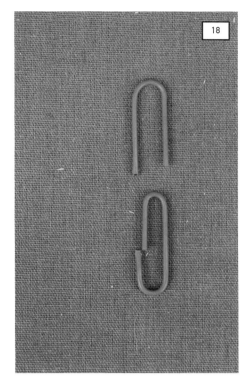

19. Glue and tape the oval section to the back-center of the base.

18. **Create the trigger.**
Cut a paper clip, as shown, or bend it back-and-forth until it separates. One section is oval, the other is shaped like a staple.

20. Bend the tip of a second paper clip, as shown.

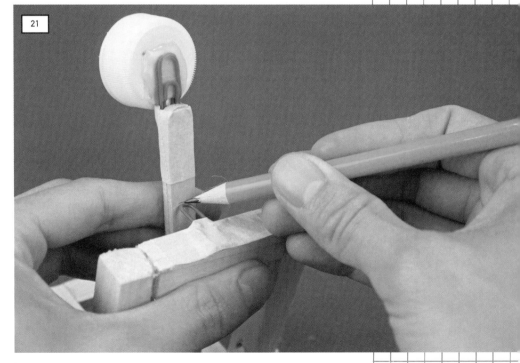

21. Pull the throwing arm back until it touches the section of paper clip that's attached to the base. Mark that spot with a pencil.

MINI MEDIEVAL SIEGE MACHINES

22 Glue and tape the second paper clip onto the throwing arm. The pencil mark lines up with the bend in the paper clip.

23 Use the staple-shaped section of paper clip for the firing pin. Straighten it out, and then bend a small handle at one end.

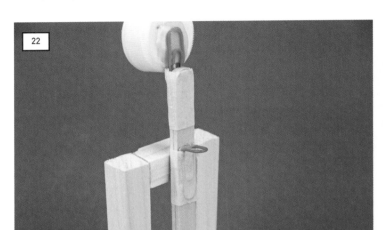

24 **Test the trigger.** Pull the throwing arm back until the paper clips line up. Insert the pin to hold the throwing arm in place.

Note:
The position of the paper clips is important. The paper clip on the throwing arm must be in **front** of the other one. It's also helpful if the tip of the firing pin is pressed against the underside of the throwing arm, as shown.

25. Get ready to fire!
Crank up the power in the torsion bundle as much as you can by removing the holding pegs and turning the winding pegs further. Remember to add tension evenly to each side. If it feels like there's so much tension that it might explode at any moment, then you're doing it right! Reinsert the holding pegs. Pull back the throwing arm and set the trigger. Load your projectile of choice (small, dense, and nondangerous objects, such as pencil erasers, work best). Take aim and pull the firing pin to shoot!

Note:
It's easier to do this step with a buddy. Have one person turn the winding pegs while the other person inserts the holding peg.

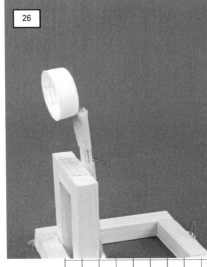

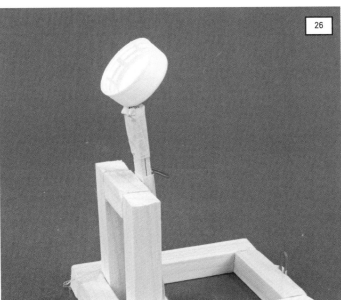

26. Adjust your trajectory.
The paper clips that connect the bottle cap to the throwing arm can be bent to adjust the angle at which your projectile is fired. If the bottle cap is facing straight forward, your projectile will also shoot forward in a straight line. If you angle the bottle cap upward, you'll fire in a high-flying arc. Experiment to find the best angle for your mini-siege needs!

Pro Tip: Long Live Your Catapult
Give your catapult a long life by letting the torsion bundle relax and unwind when you're not using it to cause mini-mayhem. You don't need to completely unwind the string, but relax it enough so that the throwing arm is pressed gently against the uprights. If you keep the torsion bundle under tension, the string will deteriorate faster, and your winding pegs may also weaken or break.

DA VINCI CATAPULT

The Da Vinci catapult, named after the famous polymath, employs two curved bows to power the catapult. But unlike a bow and arrow, the string on these bows is wound around a rotating drum. When spun, the drum pulls on the strings and bends the dual bows, storing energy inside the bending material!

This design also features an extra-long throwing arm and an adjustable basket, which give you the power to achieve some far-flung shots!

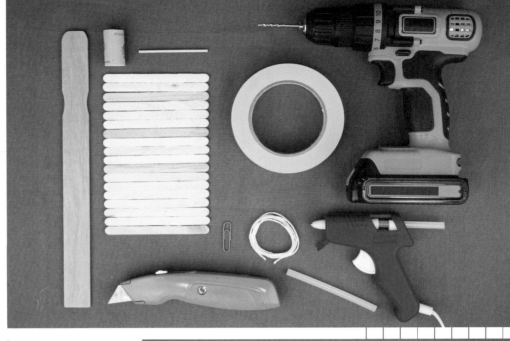

Tools and Materials

- 18 regular ½" × 4½" (1.3 × 11.4 cm) craft sticks
- hot glue and extra glue stick
- one 1" × 12" (2.5 × 30 cm) wooden paint stirrer
- utility knife
- drill and ⅛" (3 mm) drill bit
- wine cork
- ⅛" (3 mm) skewer or dowel
- cotton string
- paper clip
- masking tape
- plastic bottle cap

MATERIAL SUBSTITUTION
Wood paint stirrer: You can build the base entirely out of craft sticks, but make sure it's 2" (5 cm) wide.

1

Bend the craft sticks. Soak twelve craft sticks in warm, soapy water for at least 60 minutes. The soap in the water helps break the water tension, which will allow it to soak into the wood more easily. You will need only four bent sticks per catapult, but many will break during the bending process, so it's good to have extras.

2

Carefully bend the craft sticks into a C-shape and insert them into a straight-walled container that has an inner diameter of about 2½" (6.4 cm). Mugs are often about this size. Let the sticks dry out completely overnight inside the mugs.

3

Select four sticks that are unbroken and have an evenly curved shape. Hot glue them into pairs, as shown.

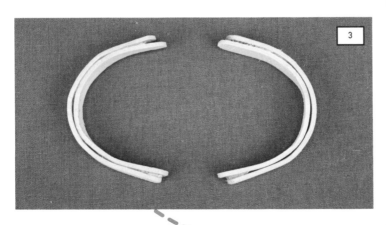

MINI MEDIEVAL SIEGE MACHINES

67

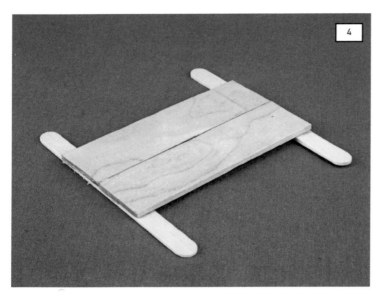

Create the catapult base.
Use a utility knife to score two 4" (10 cm) sections of a paint stirrer, then snap off the pieces. Glue two craft sticks to the undersides of the paint stirrer pieces, as shown.

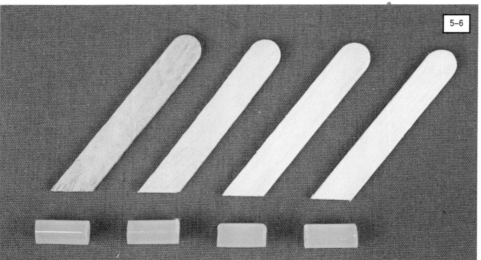

Create the supports.
Use a utility knife to score and snap one craft stick at a roughly 45-degree angle, and about 2¼" (5.7 cm) long. The precise angle and length of the cut isn't crucial, but it is important that *all* four sticks have the same angle and length. To ensure that, use the first cut piece of craft stick as a template. Lay it atop another craft stick, and use the utility knife to make another score line. Repeat two more times until you have four identical pieces, as shown.

6 Cut a mini glue stick into four ½" (1.3 cm) pieces.

7 Carefully glue the craft sticks into triangle shapes near the edges of the base. It's important that the tops of the triangles are the same height and perfectly aligned with each other.

8 Use liberal amounts of hot glue to affix the pieces of glue stick into the corners between the supports and the base. These will help prevent the supports from wobbling.

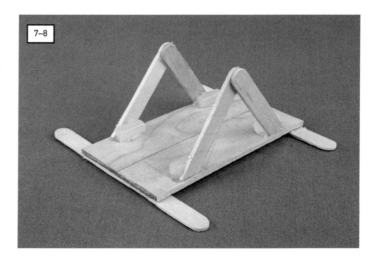

9

Create the winding drum.
Carefully drill a hole through the top of each of the supports. Hold a scrap of paint stirrer behind the support as you drill to protect your fingers.

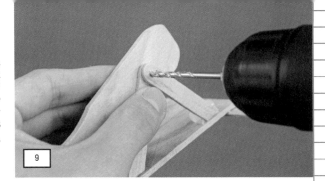

10

Carefully drill a hole through the center of the cork. As you drill, look at the drill bit from several different angles to make sure that it's going in straight. Work slowly and diligently to ensure that the drill bit exits the cork through the center.

11

Thread the dowel through the supports and cork, as shown. Apply a small amount of hot glue on the ends to prevent the dowel from sliding out.

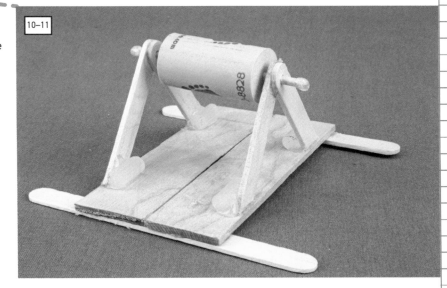

12

Create the power source.
Glue the pairs of bent craft sticks onto the base, as shown. Note that they are intentionally positioned off center.

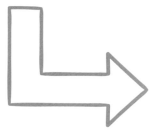

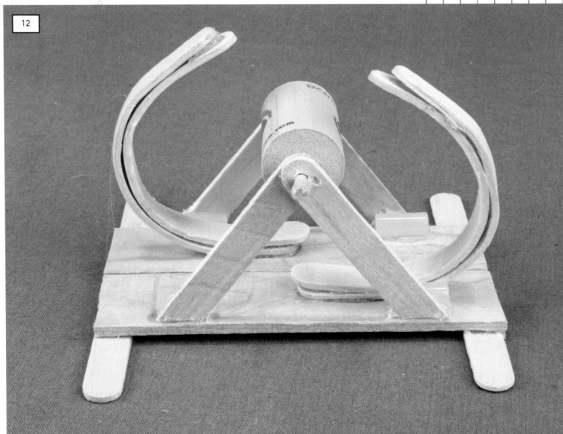

Attaching the string is the trickiest part, and it's the most important step to ensure that your catapult has maximum firing power. Read these steps very carefully.

Cut two 8" (20.3 cm) pieces of cotton string. Glue about 1½" (4 cm) of each string onto the top of each of the bent craft sticks. Wait for it to dry completely.

Each string is attached a little differently. Note that one string begins wrapping *over* the cork, and the other begins wrapping *under* the cork. It doesn't matter which one is which, but it's very important that they wrap around the cork in opposite directions. Keep this in mind for the following steps.

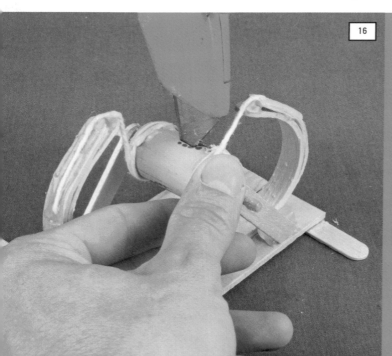

14

Wrap one string completely around the cork and pull it taut. While using one hand to hold the taut string in place, smear hot glue onto as much of the exposed string as possible, being careful not to get the glue gun too close to your string-holding hand. Patiently continue to hold the string tight while the glue dries, then let go of the string and smear more glue onto the part that was previously covered by your hand.

15

Repeat with the other string, and remember to wrap it the opposite way around the cork. While gluing on the second string, it's *very* important that first string remains tight. If the first string goes slack while you're attaching the second one, then the catapult won't be as powerful.

The photo displays the correct final result: The strings are wrapped in opposite directions, and both have equal tension.

16

Create the throwing arm. Rotate the cork so both strings are taut. Firmly pinch one side of the cork to hold it steady, and *very carefully* insert the tip of the utility cutter with the blade facing *away* from your hand. Gently wiggle the blade into the cork until you feel the dowel.

Do not try to push the blade in with excessive force. Work slowly and in a controlled manner, allowing the weight of the utility cutter to ease the blade into the cork. This step cannot be completed earlier because the incision needs to be made at the top of the cork, and that position is difficult to predict until the strings have been attached.

17

Apply a glob of hot glue over the incision. While it's hot, quickly and fully insert a sturdy craft stick. This technique will force the melted glue into the incision with the craft stick, which will firmly bond it to the cork.

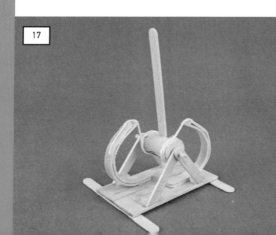

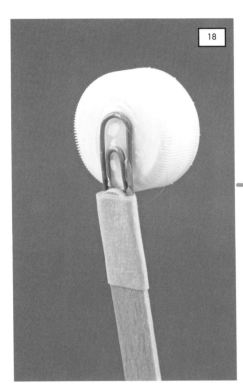

18

Glue and tape about half of a paper clip onto the end of the throwing arm. Glue the other half of the paper clip to the back of a plastic bottle cap. Keep adding layers of glue until the paper clip is completely encased.

The paper clip will allow you to adjust the angle of the bottle cap, which will give you greater control over the trajectory.

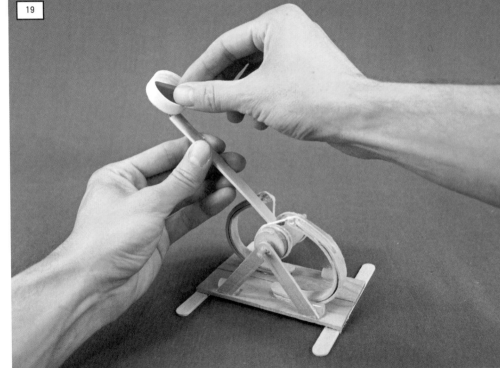

19 >>>>>>>>>>>>

Load, aim, and fire!
Use one hand to pull back the throwing arm slightly, then load your projectile of choice. As always, small and dense objects (such as a pencil eraser) work best.

Use your index finger to pull the throwing arm all the way back while the other hand holds down the front of the catapult. Note how much the craft sticks bend!

Release all the energy at once by allowing your finger to quickly slip off the side of the throwing arm!

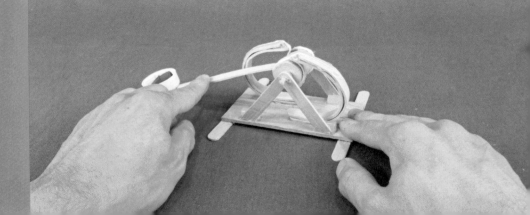

BALLISTA

Like the onager, the ballista utilizes torsion power, but with a twist: Two separate torsion springs, turned sideways, work together to hurl mini bolts—perfect for cardboard target practice! The key to a successful mini ballista is creating durable torsion bundles that can be tightly wound without breaking. Cotton string is too thick and frays too easily at such a tiny scale, so you'll be using a clever alternative: dental floss!

Tools and Materials

- nine regular 4½" × ½" (11.4 × 1.3 cm) craft sticks
- hot glue
- pliers
- drill with ⅛" and 1/16" bits (30 mm and 15 mm)
- waxed dental floss
- ⅛" (3 mm) bamboo skewers
- small paper clips
- utility knife
- small ¾" (1.9 cm) binder clip
- masking tape
- scissors

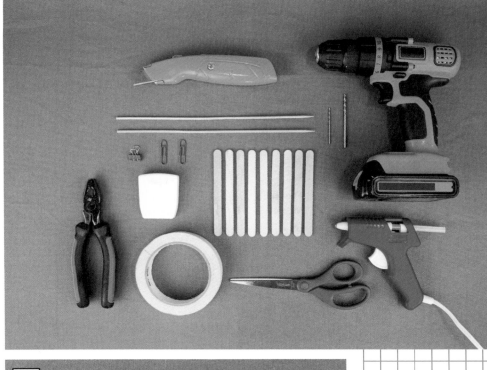

Build the frame.
Hot glue the edges of two pairs of craft sticks together, as shown.

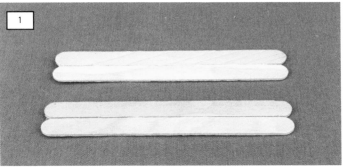

Overlap the glued pairs of sticks by about 2" (5 cm). The bottom pair of sticks will be the front end of the ballista.

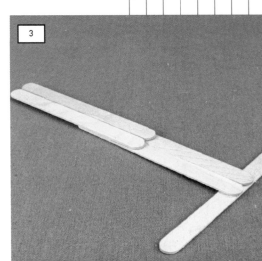

Glue a single stick onto the underside of the front of the frame.

Use a pair of pliers to snap off sixteen pieces of craft sticks, about ⅜" (1 cm) square.

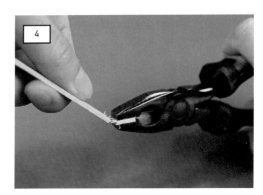

Glue the square pieces to the frame in stacks of four and then glue another stick on top. If you have a more convenient material than bits of craft sticks, such as corks or wooden cubes, use those.

Create the torsion bundle and throwing arms.

Drill two ⅛" (3.2 mm) holes, and four 1/16" (1.6 mm), as shown. Be sure all the holes are space about ¼" (6.4 mm) apart.

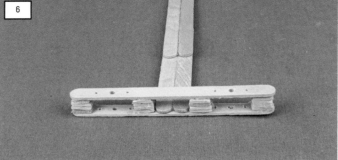

LAUNCHERS, LOBBERS, AND ROCKETS ENGINEER

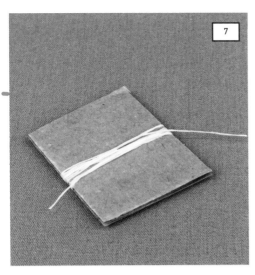

7

Cut a scrap of cardboard to about 1" (2.5 cm) wide. Wrap the waxed dental floss around it nine or ten times. Tie the ends together and remove the floss from the cardboard. Repeat to create a second loop.

8

Straighten and cut a paper clip into two 1¼" (3.2 cm) pieces, and cut one of the bamboo skewers into four 1" (2.5 cm) and two 3½" (9 cm) pieces.

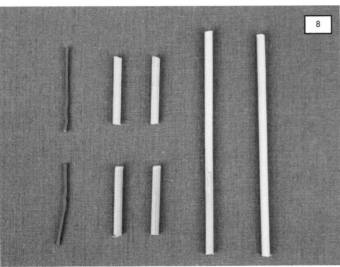

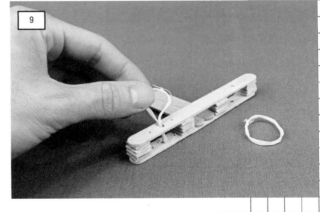

9

Select one of the loops of floss. Pinch it at one end, and insert it into one set of the ⅛" (3.2 mm) holes. Once through the holes, separate the strands, as shown.

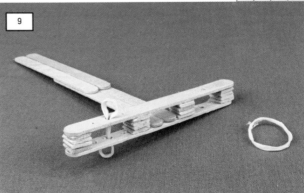

MINI MEDIEVAL SIEGE MACHINES

10

Insert the 3½" (9 cm) and two of the 1" (2.5 cm) pieces of skewer between the strands of floss, as shown. The 3½" (9 cm) skewer is the throwing arm and the two 1" (2.5 cm) skewers are now the pegs that will wind up the torsion spring. Repeat steps 9 and 10 on the other side.

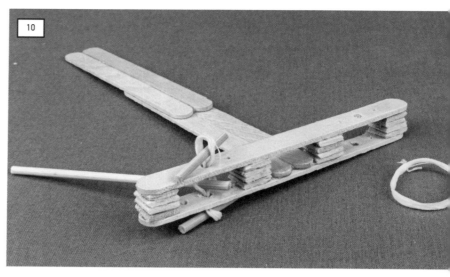

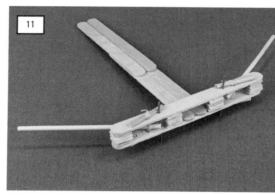

11

Rotate the 1" (2.5 cm) winding pegs to make the 3½" (9 cm) throwing arm press into the front of the ballista. In this example, the pegs are wound counterclockwise. Make sure to rotate each one evenly. When the throwing arm is pressed firmly against the frame, insert one of the paper clip pieces into the appropriate hole to prevent it from unwinding. For now, don't wind it too much—you'll increase the tension later.

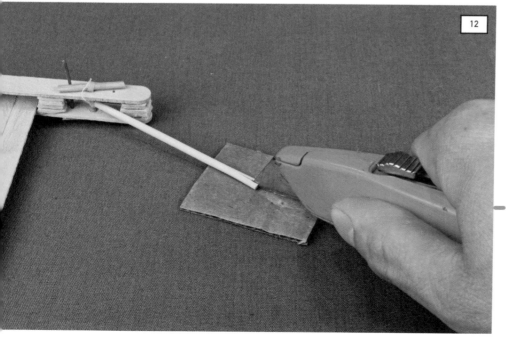

12

String the ballista.
Carefully split the tip of each throwing arm with the utility knife.

13

Slip an 11" (28 cm) piece of floss into the split, with at least 1½" (4 cm) hanging from one side. Wrap and tie the short end of the floss tightly around the tip of the throwing arm to hold it in place and prevent the throwing arm from splitting further.

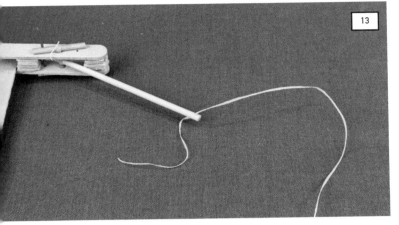

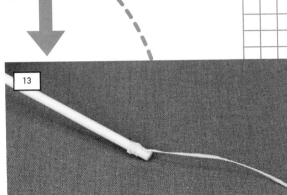

14

Repeat on the other side. Cut off the excess floss. Apply a dab of hot glue onto the wrapped floss to prevent it from unwinding.

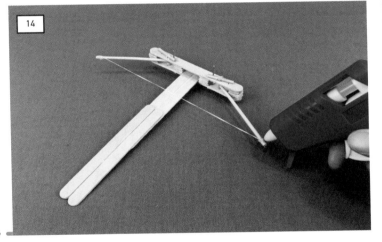

15

Create the trigger and bipod.
Finish the ballista by hot gluing a binder clip onto the back of the frame. Cut a craft stick in half at about a 45-degree angle, and glue the pieces to the front of the ballista as shown. Make sure the ends of the pieces do not protrude in front of the center gap.

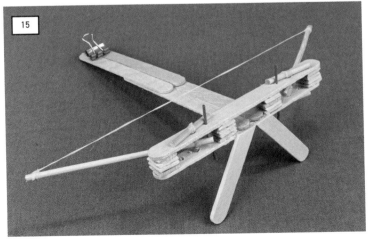

MINI MEDIEVAL SIEGE MACHINES

1

Create the bolts.
Cut the other skewer to 7" (17.8 cm), then carefully fold a 1" (2.5 cm) piece of tape over the blunt end of the skewer, leaving about ¼" (6.4 mm) exposed at the very end.

2

Trim the fins into a tapered shape that protrudes no more than ¼" (6.4 mm) from the skewer. The fins must be able to fit easily through the front gap of the ballista frame.

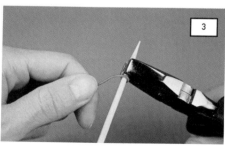

3

Straighten a small paper clip, then wrap it tightly around the tip of the skewer. This creates a leading weight (see page 19).

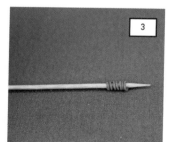

4

Use the tip of the utility knife to carefully carve out a small nock at the back of the skewer. Make sure the gap in the nock is aligned with the fins, as shown.

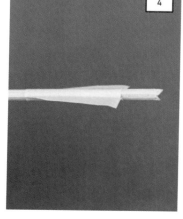

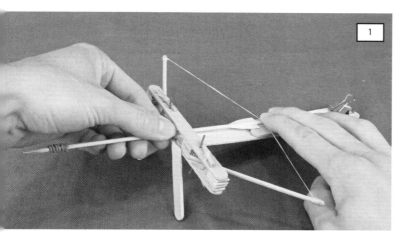

1

Get ready to fire.
Add tension to the winding pegs. Remove the paper clips and rotate the winding pegs evenly to slowly increase the tension. The bundled dental floss is extremely strong, so don't be afraid to really crank it up. (The winding pegs will break before the floss does!) Be sure to rotate the top and bottom winding pegs evenly, and check that the tension is even for both throwing arms. If one arm is more powerful than the other, your shot won't be accurate.

2

Load and fire! Insert the bolt through the front of the ballista. Position the nock on the string. With your other hand, hold the back of the bolt in place. This frees the first hand to switch positions.

3

Pinch the bolt near the fins and open the trigger. Push the bolt into the trigger.

4

Center the bolt, aim, and press the trigger to fire!

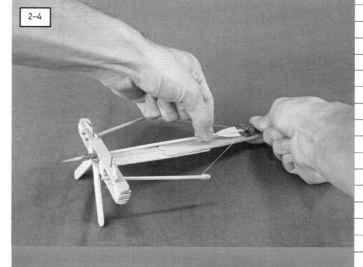

Pro Tip: Take Out the Tension
Taking out the tension will improve the longevity of your ballista. When finished firing, unwind the winding pegs one full rotation before storing it for next time.

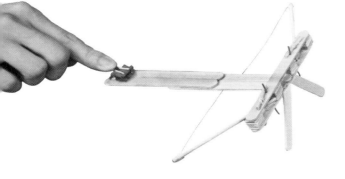

MINI MEDIEVAL SIEGE MACHINES

4 CURIOUS CONTRAPTIONS

ROLLER-AMPLIFIED MANY-THING SHOOTER

To fully harness the energy of a 7" (18 cm) rubber band, you'd need to stretch it out more than 24" (61 cm), which is quite unwieldy for a DIY sidearm! The roller-amplified shooter solves this by stretching the rubber band from the back of the frame, to the front, and then to the back again. Additionally, a small roller at the front of the frame allows the rubber band to be stretched evenly, ensuring that each shot achieves maximum power!

LAUNCHERS, LOBBERS, AND ROCKETS ENGINEER

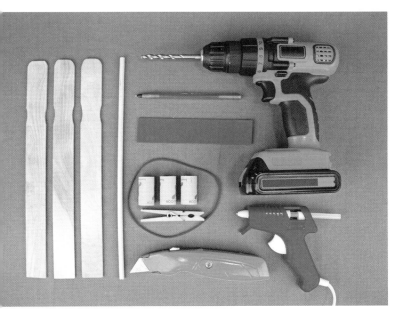

TOOLS AND MATERIALS

- three 12" × 1" (30.5 × 2.5 cm) wooden paint stirrers
- utility knife
- sandpaper (optional)
- drill with ¼" (6 mm) bit
- hot glue
- ballpoint pen casing
- ¼" (6 mm)-thick wooden dowel
- 7" × ⅛" (18 cm × 3 mm), size #117B rubber band
- wooden clothespin
- three wine corks
- craft foam, card stock, or cardboard
- another, smaller rubber band (optional)

MATERIAL SUBSTITUTIONS
Wine corks: Find another sturdy material for the handle, such as a piece of wood. Instead of cutting corks for projectiles, use pencil erasers or other dense, but soft, objects

MATERIAL REQUIREMENTS
The inner diameter of the ballpoint pen casing must be greater than the width of your dowel. If it's not, find another material that can act as a durable cylinder, such as a marker.

1 Build the frame.
Use a utility knife to score across the middle of a paint stirrer, then break it in half. Use sandpaper to clean up the edges, if desired.

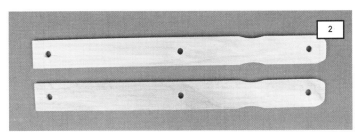

2 Stack the remaining paint stirrers. Drill three ¼" (6 mm) holes through both paint stirrers: one hole near each end and one in the middle. Allow the weight of the drill to do the work: If you push down too hard, the paint stirrer may split.

CURIOUS CONTRAPTIONS

Assemble the frame with hot glue as shown, with the holes as perfectly aligned as possible.

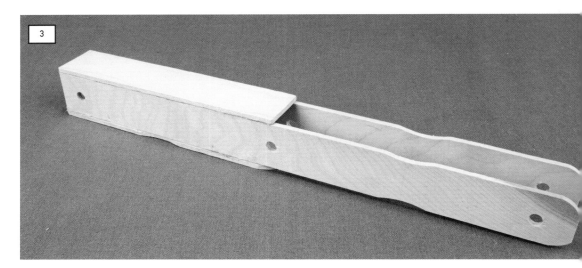

Attach the rubber band and roller.
Disassemble a ballpoint pen, then use your cutting tool to cut off a ¾" (1.9 cm) piece of the case. It's important that the case be shorter than the width between the sides of the frame.

Cut the dowel into three 1" (2.5 cm) pieces.

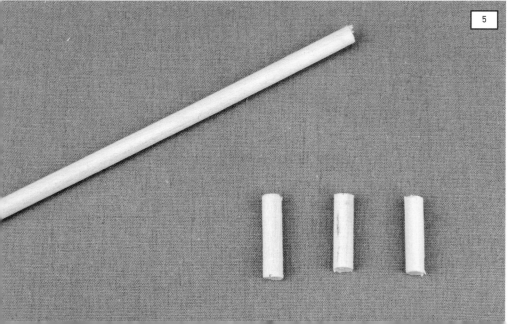

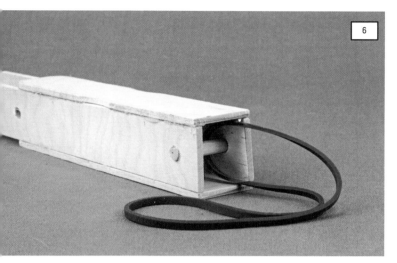

Insert the dowel through the holes, as shown. Slip the rubber band around the dowel before pushing it all the way through. This holds the rubber band firmly in place when being stretched for firing.

7

Insert another dowel piece through the holes in the middle of the frame. Again, slip the same rubber band around the dowel before pushing it all the way through. This will prevent the rubber band from getting stuck inside the frame after each shot.

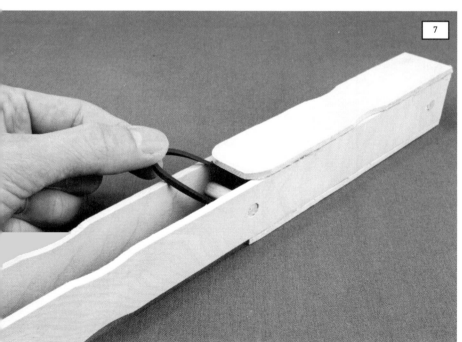

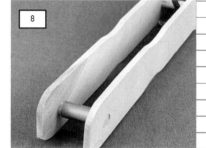

Thread the last dowel piece through one of the holes on the open end of the frame, through the piece of pen casing, and into the other hole. Check to make sure the casing can roll on the dowel without excessive friction. This is the roller that will allow the rubber band to stretch out evenly while being stretched for firing.

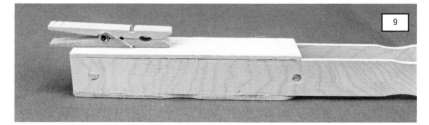

Create the trigger by gluing a clothespin to the back of the frame, as shown.

CURIOUS CONTRAPTIONS

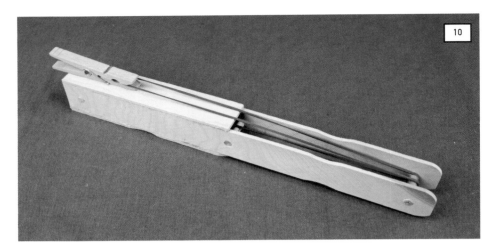

10

Test the trigger by loading the rubber band as shown. If your clothespin isn't strong enough to hold the rubber band, wrap another, smaller rubber band around it, as shown below, to increase its grip.

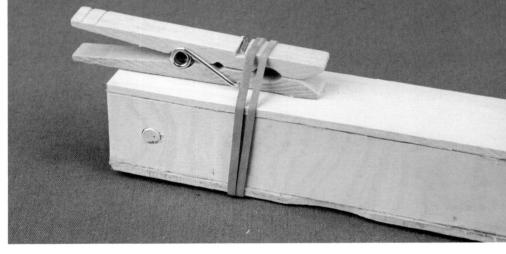

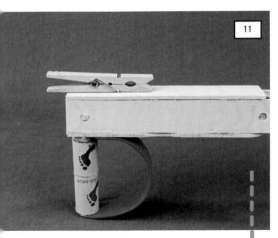

11

Create the handle and finger guard. Glue two corks end to end at the back of the frame. Glue on a 6" × 1" (15 × 2.5 cm) strip of crafting foam, as shown. The foam prevents the rubber band from slapping your fingers as it rapidly retracts! (It also contributes to the shooter's aesthetics.)

12

Create durable and safe projectiles by cutting the last wine cork into quarters. If you don't have wine corks, find something that's similar in size. The projectiles work best if they can fit snugly between the sides of the rubber band.

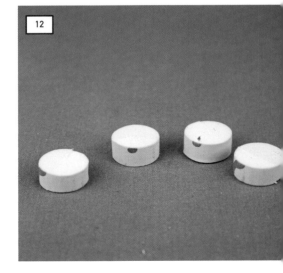

13

Prepare your shot.
Pull the rubber band under the roller at the front of the frame. Use two fingers to spread the rubber band and your other hand to open the clothespin. Push the rubber band as far into the clothespin as possible. (The clothespin's mechanical advantage is greatest near the hinge; The tip of the clothespin may not be strong enough to hold in the rubber band.)

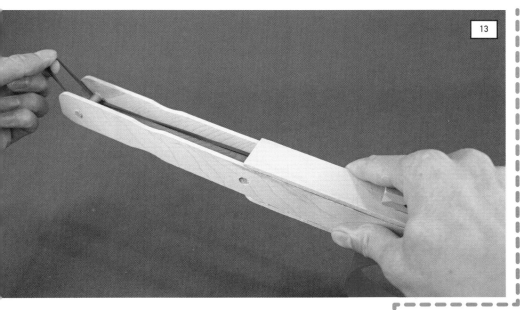

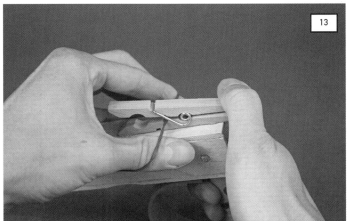

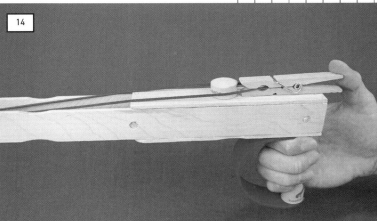

14 Set up your projectile directly in front of the clothespin. Take careful aim and fire!

Get Tinkering
This is just an example of a roller-amplified shooter. You can increase the power by making the frame longer, or experiment with different projectiles to find the best one. Build an ammo-storage compartment on the side! Or get really inventive by modifying the trigger so it's activated by your index finger instead of your thumb. This project is easy to redesign.

CURIOUS CONTRAPTIONS

DESK DRAWER BOOBY TRAP

Finally—a way to put office supplies to good use! Picture this: You're at the office, and there's someone who deserves a surprise prank. But you don't have any craft materials or fancy tools like glue guns at hand to help you pull it off. Fortunately, this desk-drawer booby trap is built using only common office supplies! When it's sprung, it releases a volley of paper clips! Best of all, this trap is specifically designed to remain armed for several days, so set it up anytime, and wait for the fun.

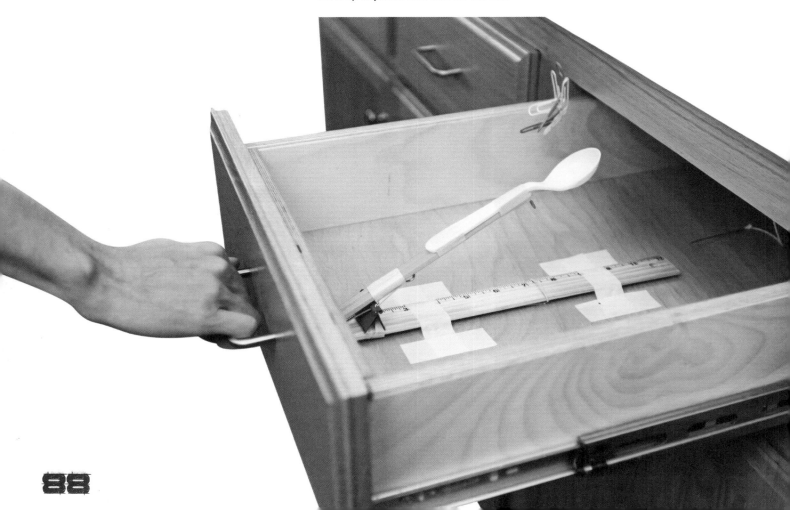

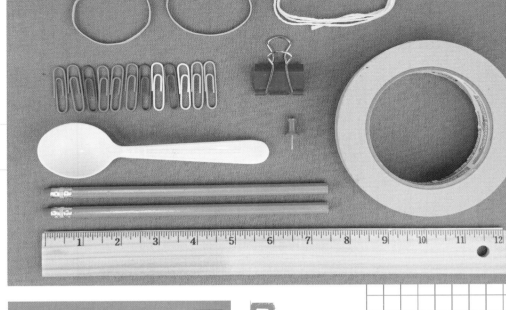

Tools and Materials

- masking tape
- 1" (2.5 cm)-wide binder clip
- ruler
- two pencils with eraser ends
- plastic spoon
- two 3½" × ⅛" (9 cm × 3 mm) size #33 rubber bands
- small handful of paper clips
- thumbtack
- string

MATERIAL SUBSTITUTIONS
Wooden measuring stick: Any flat and strong object, such as a paint stirrer

#2 pencils: Disposable wooden chopsticks or any other rigid, flat stick

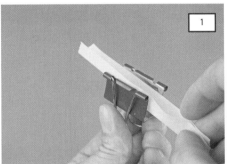

1 Create the catapult. Cut a 4" to 5" (10 to 13 cm) piece of tape and crease it in half, lengthwise, sticky-side out. Open the binder clip and place the piece of tape inside.

2 Position the binder clip about 1½" (4 cm) from the end of the ruler. Wrap the tape around the ruler to secure the clip into place. Open the binder clip handles. This will be the catapult's fulcrum.

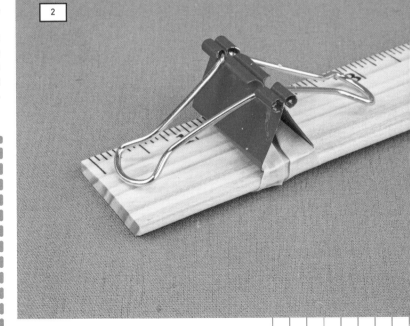

CURIOUS CONTRAPTIONS

Tape the two pencils together. Position the pencils over the binder clip so that the eraser ends line up with the end of the ruler. Tape the pencils to the inward-facing binder clip handle. Use tape to attach the plastic spoon to the other end of the pencils in two places.

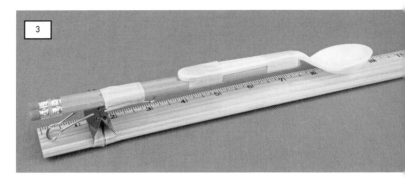

Wrap both rubber bands around the end of the ruler and the eraser-end of the pencils. Two or three wraps is sufficient.

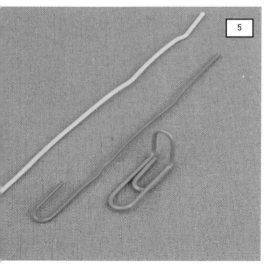

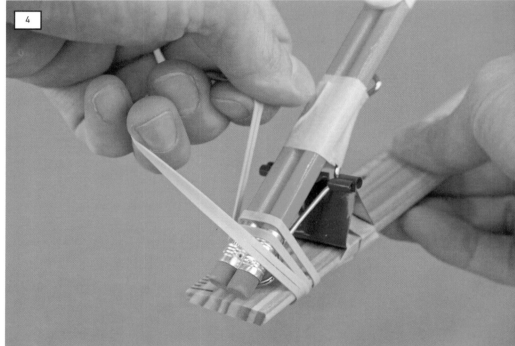

Create the trigger.
Modify three paper clips, as shown. The blue one has about ⅜" (1 cm) of one end bent upward at a 90-degree angle.

Use several layers of tape to attach the bent blue paper clip to the underside of the pencils, as shown. Tightly wrap the tape as you apply it. Layering the tape and wrapping it tightly will ensure that this part of the trigger does not come loose when the trap is armed over a long period of time.

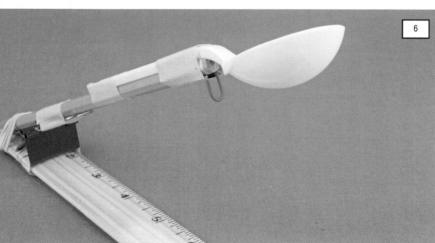

LAUNCHERS, LOBBERS, AND ROCKETS ENGINEER

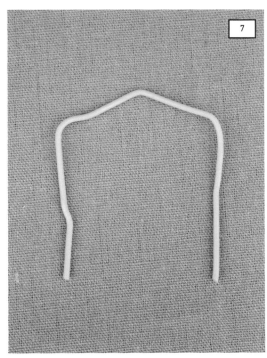

Bend the straightened yellow paper clip into a house-like shape. Make sure the "roof" is peaked. This will make it easier to arm the trap in the next steps.

Bend the yellow paper clip around the ruler, just behind the blue paper clip. Wrap the ends of the yellow paper clip around the underside of the ruler and firmly attach it in place with a piece of tape.

9

Tie the ends of a 10" (25 cm) piece of string to the red paper clip and the thumbtack as shown. Double knot to make sure they can't come loose.

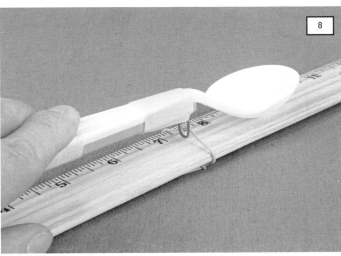

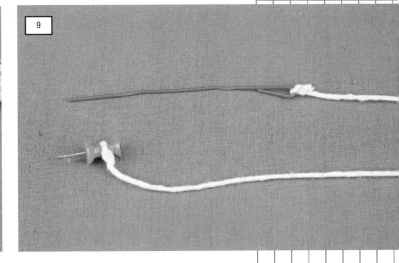

Test the trigger.
Pull the catapult all the way down, and insert the red trigger pin through the yellow and blue paper clips. Make sure it stays put. Give the string a tug to release!

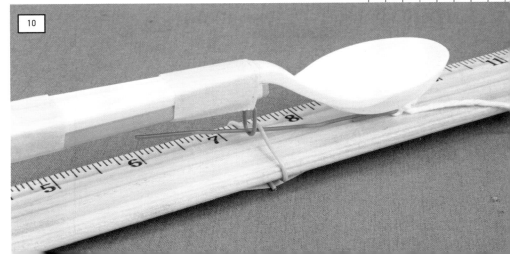

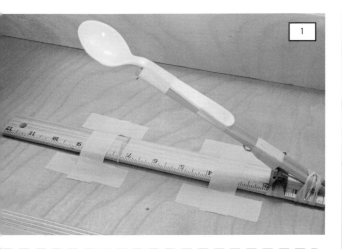

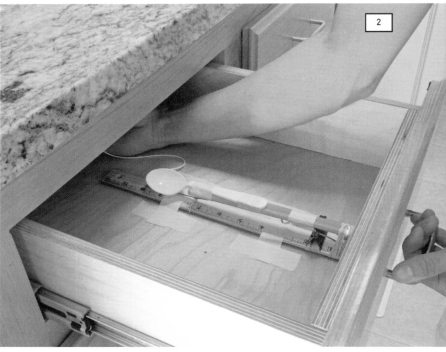

1
Arm the trap.
Set the fulcrum-end of the catapult against the very front of the drawer. Apply tape across the ruler in two places, then apply four more pieces of tape, as shown, covering the ends of the first two pieces. This taping technique will firmly secure the ruler to the bottom of the drawer.

2
Set up the catapult trigger.
Push in the drawer as far as is possible while making sure the catapult arm doesn't hit the top of the drawer.

While holding the drawer open in that position, pull the string taut and firmly push the thumbtack into the underside of the drawer.*

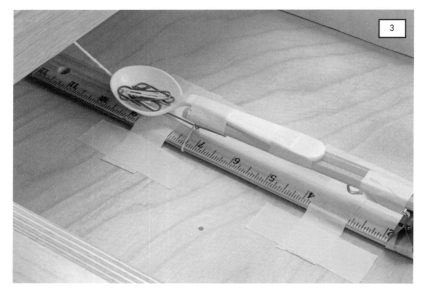

3
Give it a test: Close the drawer completely, then open it with a normal amount of force. As the drawer opens it should pull the trigger pin free, releasing the catapult!

Load the spoon with some paper clips or other small objects. Be sure to use many small (and harmless) projectiles for a maximum chance of hitting your target. Refill the drawer with its usual contents, close it, and wait for your victim to reach for the stapler.

LAUNCHERS, LOBBERS, AND ROCKETS ENGINEER

*Can't use a thumbtack?
A thumbtack won't work if the drawer is made of metal. Use this taping technique to attach the string to the underside of the drawer.

1
Apply two layers of tape to the string to affix it to the underside of the drawer. Leave about 1" (2.5 cm) of string poking out one side.

2
Fold the string over, and apply a second piece of tape. This will prevent the string from slipping.

3
Apply two more pieces of tape on each side, as shown.

CURIOUS CONTRAPTIONS

POKER CHIP SHOOTER

The vision: A clever contraption that can fire off mini frisbees, one at a time, with a high degree of accuracy, consistency, and at a distance of at least 15' (4.5 m).

The result: The Poker Chip Shooter! Though this project requires an extra degree of precision in its construction, it's 100 percent worth it to experience the gratifying action of rapid-firing poker chips one after another!

Tools and Materials

- four 12" (30.5 cm) paint stirrers that are at least 2¼" (5.7 cm) wide
- utility knife, or other tool for cutting craft sticks, paint stirrers, and corks
- ten 4½" (11.4 cm)-long craft sticks
- hot glue gun
- two 12" (30.5 cm)-long, ¼" (6.4 mm)-wide wooden dowels
- standard-size, clay composite poker chips*
- 4" (10.2 cm) toilet paper tube
- duct tape
- 3 synthetic wine corks
- four 3½" × ⅛" (9 cm × 3 mm) rubber bands

*I recommend that you use 11.5-gram clay composite chips. The thin, plastic chips and the heavyweight 14+ gram chips will not shoot as far.

This project requires precision.
This project works by carefully depositing one poker chip at a time in front of the firing mechanism. It's essential that the path of the poker chip is sized correctly and is free of even the smallest obstructions. Follow these directions carefully or you may end up with a fancy poker chip storage container rather than a sharpshooting contraption!

MATERIAL SUBSTITUTIONS
Paint stirrers: Use balsa or another thin, easily workable wood. It must be rigid and flat.

Synthetic corks: You can use regular corks, but they are slightly more prone to breaking. You can also use any other comfortable material for the handle, and rubber cabinet bumpers or another resilient material for step 12.

Poker chips: You can modify this project to work with other disk-shaped materials, such as checker pieces, backgammon pieces, or plastic pog slammers. (Anyone remember those?) If you do, you'll need to modify the dimensions of your shooter and ammo container to accommodate the size of your ammunition.

1
Build the foundation.
Line up two paint stirrers side by side and glue a craft stick onto one end, as shown. Cut four craft sticks in half. Glue three of the halves onto the paint stirrers, evenly spaced, as shown. (You'll use the other craft-stick halves later).

2
Flip the foundation over. Hot-glue a dowel along each side, as shown. Make sure the dowels are perfectly parallel to each other and that a poker chip can slide easily between them. Make sure there's no excess hot glue that might interfere with the poker chip's ability to slide. There should be only $1/16$" to $1/8$" (1.5 to 3 mm) of space between the chip and the sides at any given point.

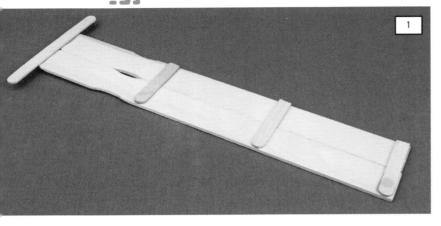

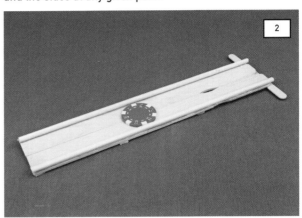

3
Build the top of the firing slot. Cut a paint stirrer in half, crosswise. Place the pieces side by side, as shown, and hot glue two of the craft-stick halves at the ends to connect them.

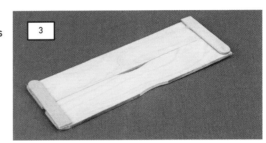

4
Line up the firing slot top along one end of the foundation. Apply several dots of glue along the dowels along the length of the top piece. Carefully attach the top to the foundation. Again, make sure that a poker chip can easily slide through the firing slot without getting caught on any excess hot glue.

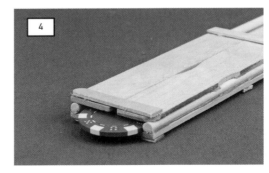

5
Glue three more craft-stick halves across the top of the dowels, as shown. These extra sticks will help keep the firing mechanism aligned with the firing slot. Make sure that the toilet paper tube (ammo container) fits snugly in the spot behind the firing slot.

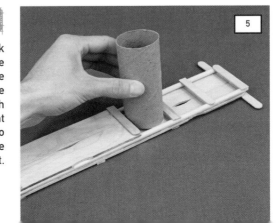

LAUNCHERS, LOBBERS, AND ROCKETS ENGINEER

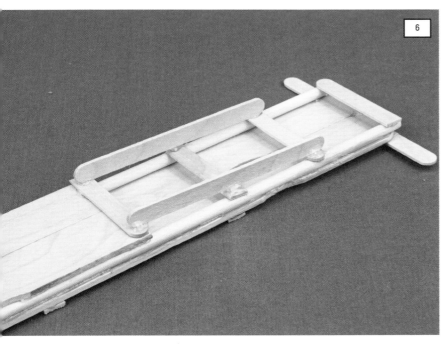

6 Glue two craft sticks on edge, as shown. These will help support the ammo container.

7 **Make the ammo container.** Wrap the toilet paper tube with duct tape to increase its durability. Make sure the tape does not extend beyond the edges of the tube or it may interfere with the firing mechanism.

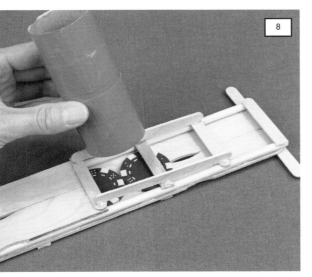

8 Line up two poker chips behind the firing slot, as shown. This will ensure that the ammo container is positioned above the foundation with just enough space to allow only one poker chip at a time to be released.

9 Rest the ammo container gently on the poker chips. Squeeze the tube to create a gap between it and the craft sticks that were added in step 6. Apply a glob of hot glue and then release the tube. Repeat on the other side.

CURIOUS CONTRAPTIONS

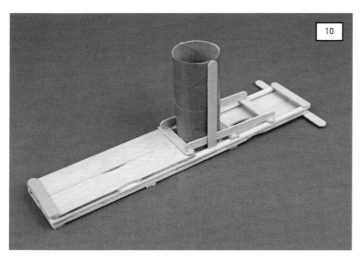

10 Glue two more craft sticks upright, on either side of the ammo container, as shown, to further secure its position.

Build the firing mechanism.
Start by gluing two craft-stick halves onto the end of a paint stirrer. Insert the paint stirrer through the firing slot, as shown. Cut two ¼" (65 mm)-thick disks of cork and glue them onto the ends of the dowels and abutting surfaces. The cork bumpers will absorb some of the impact on the firing mechanism.

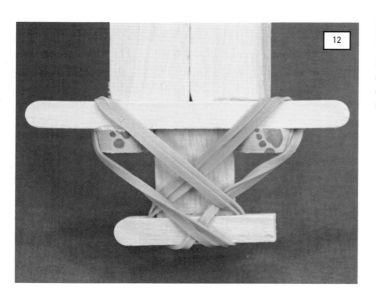

Flip the shooter over. Wrap two pairs of rubber bands in an X configuration around the whole craft stick and then around the craft-stick halves and the end of the firing mechanism. Turn the shooter right side up.

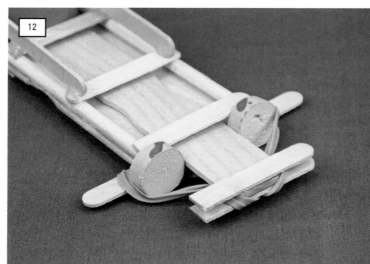

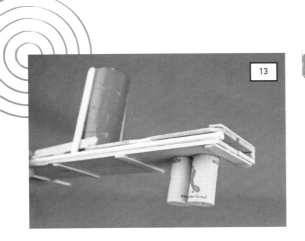

 Add the handle.
Center and glue two corks under the firing slot of the shooter, as shown. Use plenty of glue. Wait for it to dry completely before testing it.

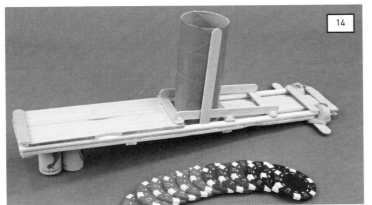

 Get ready to fire!
Fill the ammo container with poker chips, making sure that they stack flat. Grasp the handle in one hand and pinch the back of the firing mechanism with the other.

Take careful note of this next step: Pull back the firing mechanism until it's just behind the ammo container **and not any farther!** A single poker chip should fall into position under the ammo container. Release the firing mechanism and watch a poker chip fly at high velocity!

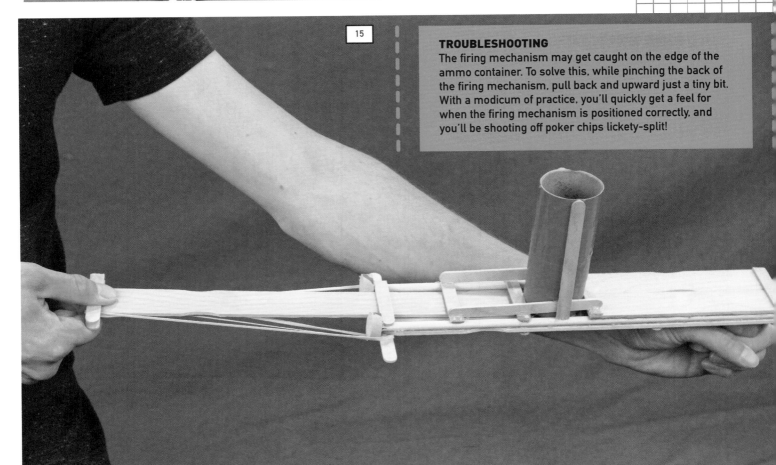

TROUBLESHOOTING
The firing mechanism may get caught on the edge of the ammo container. To solve this, while pinching the back of the firing mechanism, pull back and upward just a tiny bit. With a modicum of practice, you'll quickly get a feel for when the firing mechanism is positioned correctly, and you'll be shooting off poker chips lickety-split!

SLIDE-ACTION RUBBER BAND GUN

Sure, it's satisfying to shoot off a rubber band with the snap of a clothespin release. But it's way more fun to fire off a burst of five rubber bands at once with a quick slide-action pull on the trigger! This project is relatively simple to build and powerful enough to knock over a pyramid of plastic cups from 7' (2 m) away!

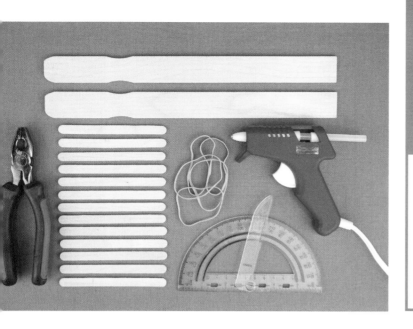

Tools and Materials

- 13 regular 4½" × ⅜" (11.4 × 1 cm) craft sticks
- pliers
- two 1" × 12" (2.5 × 30.5 cm) wooden paint stirrers
- hot glue
- protractor (optional)
- 3½" × ⅛" (9 cm × 3 mm) size #33 rubber bands

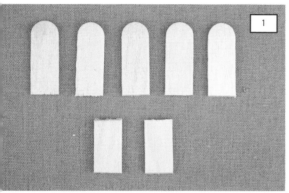

Use the pliers to break off five 1" (2.5 cm) pieces and two ¾" (1.9 cm) pieces from the ends of craft sticks. (Use the same craft-stick breaking technique as shown on page 74.)

Create the loading notches.
Glue the 1" (2.5 cm) pieces of the craft sticks onto one side of a paint stirrer, as shown. The craft-stick pieces are spaced about ³⁄₁₆" (5 mm) apart and protrude above the top edge of the stirrer by exactly ¼" (6 mm). It's important that the rounded ends of the loading notches are positioned, as shown. If they protrude too much or too little, the rubber bands won't load and release easily. Glue on two full craft sticks, as shown, to start the handle.

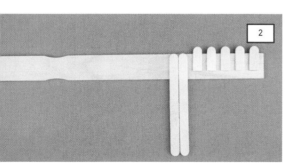

Flip the piece over. Center and glue a craft stick onto the handle. Use a second paint stirrer to determine the space between craft stick and the first paint stirrer. This will help guide the slide-action trigger and prevent it from rubbing against your hand.

Glue one of the ¾" (1.9 cm) pieces of craft stick onto the bottom of the handle.

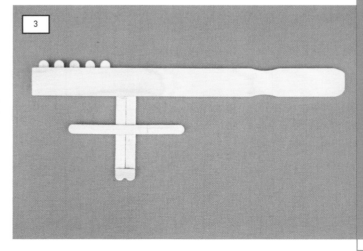

Note: Measurements May Vary
The paint stirrers used in this example are ⅛" (3 mm) thick. If yours are thinner, even by ¹⁄₁₆" (1.5 mm), then you may need to layer two craft sticks together in step 3 to create a thicker handle that can accommodate the additional thickness of the paint stirrers.

CURIOUS CONTRAPTIONS

4 **Create the slide-action trigger.**
First, make sure there is no extra glue spilling onto the area where the trigger will go. If there is, scrape it off, otherwise it will make the trigger difficult to operate.

Position a second paint stirrer as shown, but **do not** glue it down. Make sure it can slide easily between the first paint stirrer and the craft stick.

Use a protractor to mark a 35-degree angle from the bottom-left corner of the paint stirrer, then glue a craft stick along that line. Glue the stick **only** to the lower paint stirrer. This stick will lift and release the rubber bands as the trigger slides.

If you don't have a protractor, then glue the craft stick at an angle so that one end protrudes about ⅛" to 3⁄16" (3 to 5 mm) **above** the loading notches.

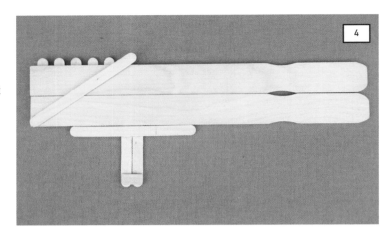

5 **Complete the handle.**
Temporarily remove the trigger, and apply glue in three places, as shown. Make sure you don't use excessive glue that might interfere with the trigger movement. Glue on two more craft sticks, as shown. Once the glue has dried, reinsert the trigger. It should slide easily through the handle.

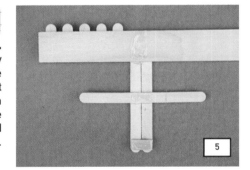
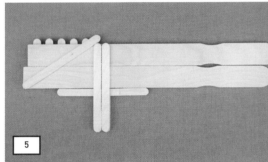

Note: Measurements May Vary
Many paint stirrers are about 1" (2.5 cm) wide, however, if yours are wider or narrower, then the angle of the trigger will be different than 35 degrees. The key is to ensure that it's protruding above the loading notches at a shallow angle. Adjust your creation accordingly.

6

Create another handle in front of the slide-action trigger. Use four craft sticks: two glued to one side, two glued to the other side. Connect them at the bottom with the remaining ¾" (1.9 cm) piece of craft stick. Make sure these are glued **only** to the slide-action trigger and not the first piece.

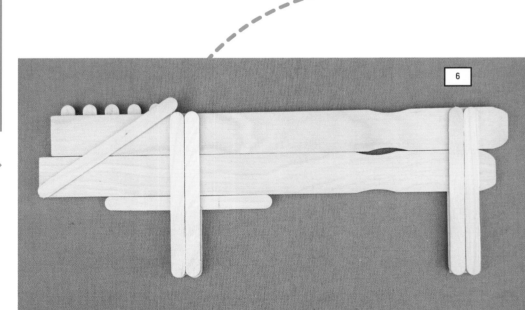

7

Create slide-stops.
This step will prevent the slide-action trigger from extending too far forward or backward. This has no impact on the performance, but it does make it more satisfying to operate.

Snap off two more ¾" (1.9 cm) pieces of craft stick. With the trigger positioned, as shown, glue one piece onto the trigger right behind the handle. Glue the other about 4" (10.2 cm) in front of the handle.

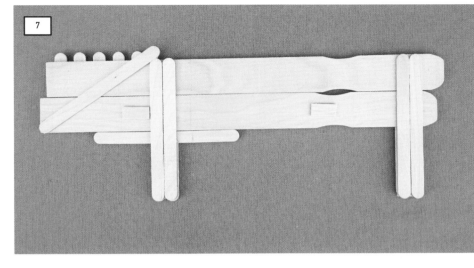

The trigger should be able to be withdrawn, as shown, but stopped by the piece of craft stick. This creates a snappy *clack* sound whenever you fire or reset the trigger!

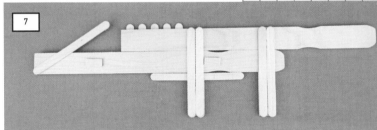

8

Load and fire!
Load the rubber bands from **back to front**. If you load the front ones first, the gun will jam. Stretch the rubber band from the upper-front tip of the gun onto the backmost loading notch. Repeat until five rubber bands have been loaded from back to front.

To fire: For a single and powerful blast, rapidly pull the slide-action trigger to release all five at once! For smaller bursts, pull the trigger slowly and deliberately.

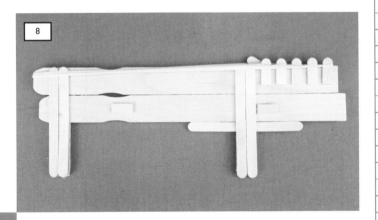

TROUBLESHOOTING
If the rubber bands aren't releasing, make sure the loading notches aren't protruding too high, and make sure that the end of the trigger stick is at least ³⁄₁₆" (5 mm) above the loading notches.

If the slide-action trigger is hard to pull, you may need to disassemble the handle and look for loose bits of hot glue that may be creating excessive friction.

Design Variables
Here are a few quick ideas to customize your rubber band blaster:

- Add more than five loading notches
- Space the loading notches more than ¼" (6 mm). This makes it easier to fire one shot at a time
- Make it longer! Stretch out the rubber bands more to get more firepower!

5 FIREARMS

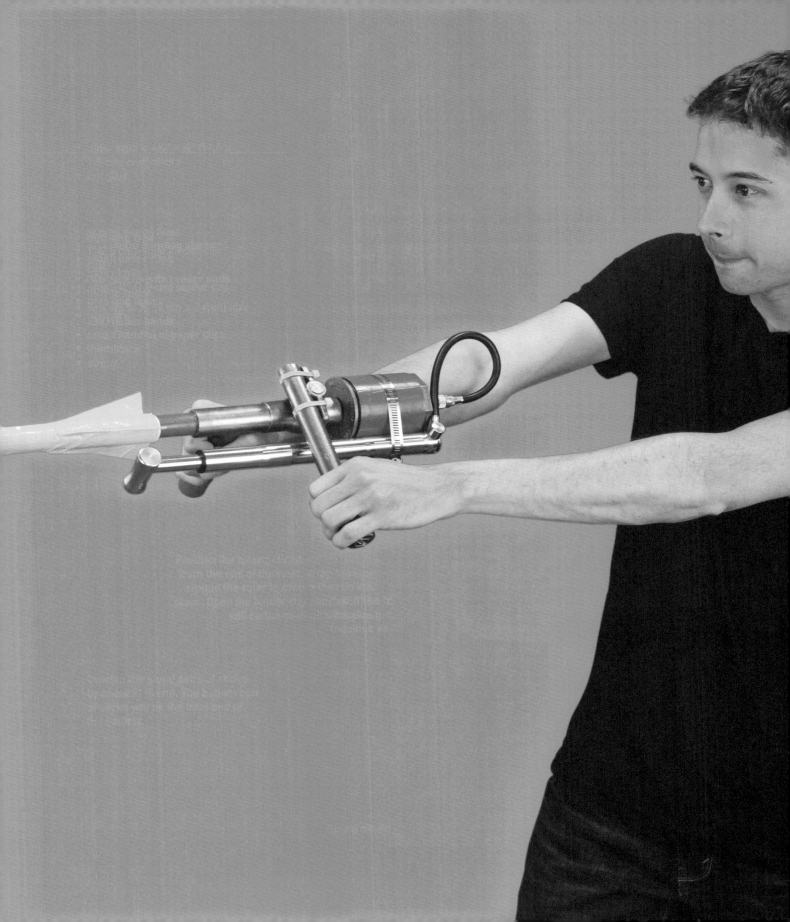

BBQ BLASTER

Transform a BBQ lighter, pill bottle, and pen into a mini combustion-based blaster! The fuel and spark from the BBQ lighter is reconfigured to produce a tiny explosion. The resulting expanding gas shoots anything that's inside the barrel at high speeds! This tiny firearm can fire a pen nib more than 20' (6 m), and it's easy to adapt it to shoot copper BBs or anything else that's less than ¼" (6 mm) wide.

Tools and Materials

- cheap ballpoint pen that can be disassembled
- wire cutters
- drill with ⅜" bit and ¹⁄₁₆" bit (1 cm and 1.6 mm)
- paper clip
- small medicine bottle with screw-on cap
- hot glue
- cheap BBQ lighter
- cable tie
- weak ferrous magnet and copper BBs (optional)

MATERIAL SUBSTITUTION
Medicine bottle: You can use any small, nonglass container with a screw-on cap, but make sure it's fairly small. Larger containers will require more time and fuel per shot. If possible, use a container that's transparent. It's quite gratifying to see the small explosion inside the combustion chamber!

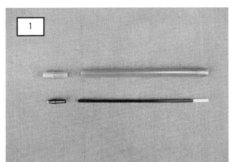

1 Create the barrel. Disassemble the pen, including the writing nib—you may need to wipe up some ink. Check that the nib can fit inside the pen casing (a.k.a. the blaster barrel).

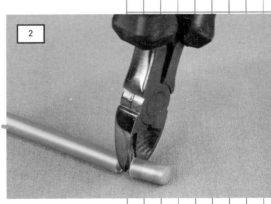

2 Use the wire cutters to chop off the closed end of the barrel.

3 Cut a paper clip in half. Drill two ¹⁄₁₆" (1.6 mm) holes into one end of the barrel, and thread the paper clip through. Bend the paper clip ends around the barrel. This will prevent your projectile from falling through.

 4
Drill a ⅜" (1 cm) hole into the center of the medicine bottle cap and insert the barrel, as shown. Make sure the paper clip side is on the inside of the cap. Apply liberal amounts of hot glue on both sides of the hole to ensure a strong seal.

 5

Modify the lighter.
Carefully open the BBQ lighter (cheaper designs tend to be easier to pull apart). Remove the lighter stem and expose the ignition wires and fuel line. Keep everything else in place (the trigger, fuel reserve, and ignition mechanism should remain in their original positions).

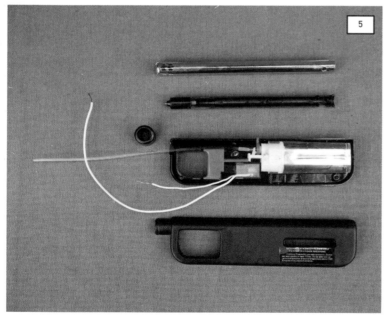

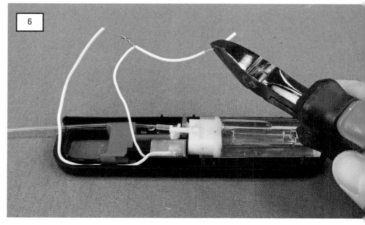

6 Many BBQ lighters have wires of different lengths. Modify the wires so they're each about the same length by cutting a piece of wire from the longer one and twisting it together with the shorter wire. You can strip the wire by carefully pinching the insulation with the wire cutters and pulling the wire through, as shown.

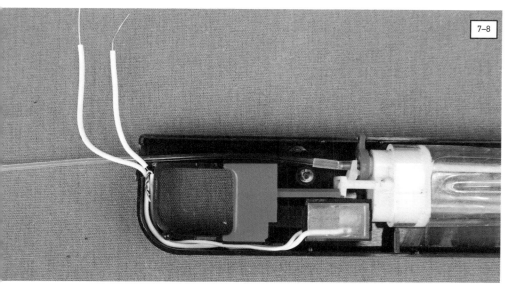

7 Tuck the wires into the lighter casing. If the casing is made of metal, cover any exposed wire with tape.

8 Test your wiring.
Position the ends of the wires so they're about ¼" (6.4 mm) apart. Click the trigger. You should see a small white spark appear. If not, try moving the wires slightly closer together or slightly farther apart. If it's still not working, check that the wires are connected to the ignition mechanism in the position you found them. If you had to cut the wire in step 6, check that your wire connections are tightly twisted together. Make sure that no metal is touching the exposed wire.

9 Reassemble the lighter with the wires and fuel line coming through the hole left by the lighter stem you removed. Use hot glue, if necessary.

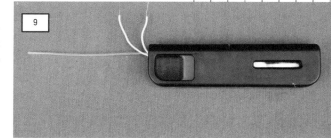

10 Attach the combustion chamber (medicine bottle). Apply liberal amounts of hot glue to the lighter, and adhere the back half of the combustion chamber to it, as shown. Hot glue does not adhere very well to some plastics, so you may need to supplement the bond by strapping the combustion chamber to the lighter with a cable tie.

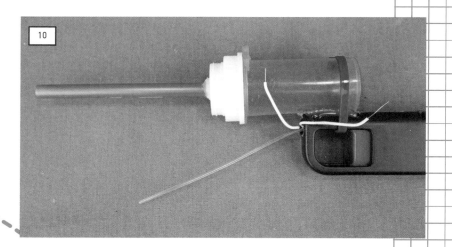

FIREARMS

11. Install the ignition and fuel line.

Drill two 1/16" (1.6 mm) holes in the **back** of the combustion chamber on either side of the lighter. Insert one wire into each hole, and position the exposed metal ends about 1/4" (6.4 mm) apart.

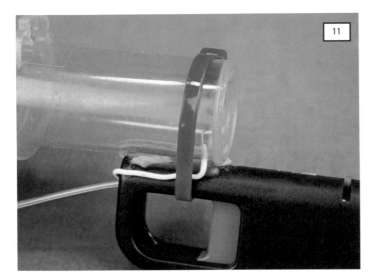
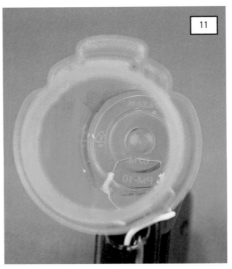

12.

Test the ignition by clicking the trigger. You should see a small white spark appear between the two wires. If not, move the wires slightly closer together or slightly farther apart. When you've found a reliable position, apply hot glue on the outside of the wire holes to keep the wires in place.

13.

Drill a final 1/16" (1.6 mm) hole in the **front** of the combustion chamber, and insert the tip of the fuel line. If it's a snug fit, don't apply hot glue or you may melt the plastic tubing. If it's loose, apply a small bead of hot glue next to the fuel line, wait a few seconds, and then smear the glue around the tubing with a scrap of cardboard.

Why is the ignition in back and the fuel in front?
Combustion occurs only when the right mixture of oxygen and fuel meets a spark. If the fuel line is too close to the ignition, it will smother all the surrounding oxygen, and combustion won't occur. Additionally, positioning the ignition in the back of the medicine bottle ensures that the combustion starts there, and rapidly expands toward the barrel, resulting in a much more effective explosion.

Load and fire!
Drop the pen nib into the barrel. Make sure the metal tip of the nib is facing outward. Carefully take aim.

Safety First
Despite its small size, the BBQ Blaster can fire the pen nib with surprising speed. Take aim at cardboard or other absorbent surfaces. *Do not* fire at walls or other hard surfaces or the pen nib may ricochet back at you!

Always wear safety goggles while firing.

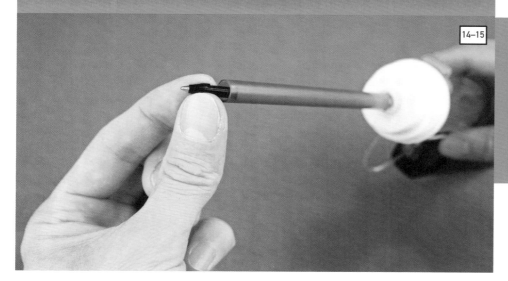

Begin filling the combustion chamber with fuel by squeezing the trigger slightly (Most BBQ lighters begin releasing fuel when the trigger is pulled part way). After 1 or 2 seconds, click the trigger repeatedly. You'll see a burst of light fill the chamber, hear an airy POP, and see the pen nib zip through the air!

Reset
After each shot, unscrew the combustion chamber and blow some air inside to clear out the fumes. Combustion needs plenty of oxygen, so it won't work unless you clear out the fumes and introduce fresh O_2.

You may be tempted to try filling the blaster with fuel for more than two seconds to get a more powerful shot . . . sorry to tell you, this probably won't work! Remember that combustion needs oxygen. If you fill the entire combustion chamber with fuel, there won't be enough oxygen for ignition to occur.

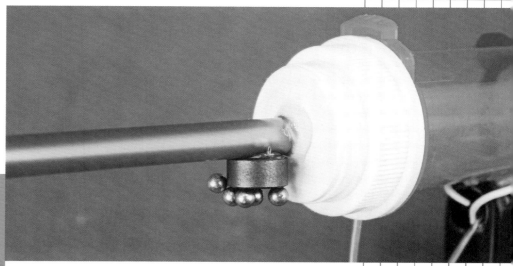

Optional: Make It a BB Blaster
If you can't find a pen with the right-size nib, or if you want to have a bountiful supply of alternate ammo, you can modify the blaster to shoot BBs! Simply hot glue a weak ferrite magnet onto the underside of the barrel, as shown. The magnet will prevent the BB from rolling out, but it should be weak enough not to hinder the blaster's performance. This also allows you to store extra ammo on the outside of the magnet!

Drop a copper BB into the barrel and fire as usual.

FIREARMS

PING-PONG BALL MORTAR

Of all the projects in this book, the Ping-Pong Ball Mortar may have the best time-and-effort-to-gratification ratio. The construction is straightforward, and the fireball-like combustion reaction can fire a ping-pong ball more than 40' (12 m)—remarkably far for such a lightweight object! You'll get tons of explosive excitement without needing to worry about accidentally dropping a bombshell on your car!

TOOLS AND MATERIALS

- 4 paper towel tubes
- utility knife (or other tool to cut cardboard & plastic)
- plastic bottle with screw-on cap, and diameter of 2" (5 cm) or less
- duct tape (2 colors optional)
- drill with ⅜" (1 cm) bit
- ping-pong ball
- aerosol hair spray
- BBQ lighter

MATERIAL SUBSTITUTION
Paper towel tubes: If you don't have four on hand, you can get by with two. Use strips of cardboard or another material for the bipod that holds the barrel upright.

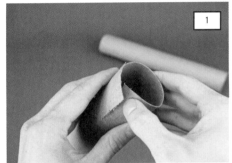

1

Make the barrel. Cut a 2" (5 cm) slit into one of the tubes and fold the edges together slightly.

2

Insert the tube into another one.

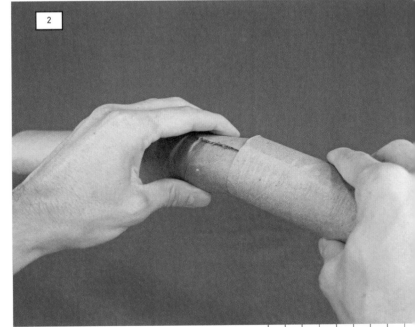

FIREARMS

3

Carefully cut off the bottom of the plastic bottle.

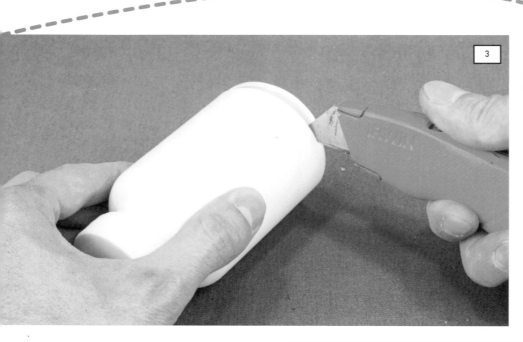

4

Slip the bottle over one end of the tubes and then cover the whole thing with long strips of duct tape. (Optional) Wrap the ends in red-colored tape for a snappy look.

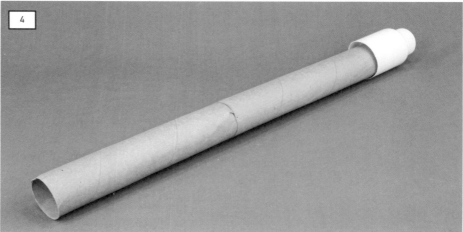

5

Create the bipod.
Apply more long strips of tape along the length of the remaining tubes. The tape isn't necessary, but it does add durability and makes the bipod look much better.

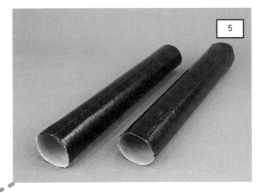

Pinch one end of each tube and seal it closed with more tape.

7

Tape the pinched end of each tube onto the side of the barrel, about a quarter of the way down from the end.

The angle at which you attach the bipod legs is up to you. For a higher shot, tape them in a way that angles the barrel upward. To transform this into a cannon-like launcher, angle the barrel upward only slightly.

8

Create the ignition site.
Drill a hole into the top of the plastic bottle near the neck. Make sure the hole is wide enough so that the stem of the BBQ lighter can fit snugly inside.

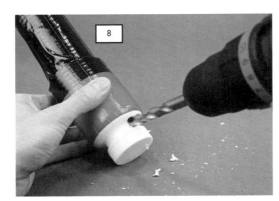

FIREARMS

9 Load and fire!
Drop the ball into the barrel. The ball should stop somewhere in the middle of the barrel because of the cut-and-folded section created in step 1. If it doesn't, see Pro Tips and Troubleshooting.

Position the mortar **first**. If you take too long to move and aim, the fuel will evaporate.

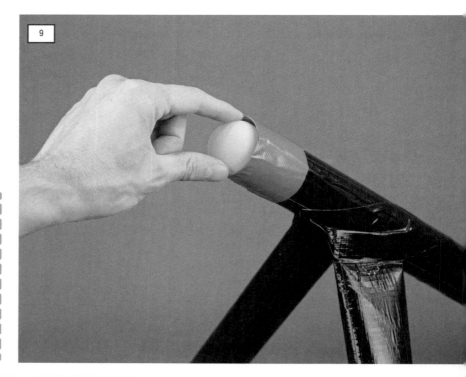

10 Unscrew the bottle cap and fill the back half of the barrel with a 1-second burst of hair spray. Don't overfill it or it won't work: There needs to be enough oxygen for combustion to occur.

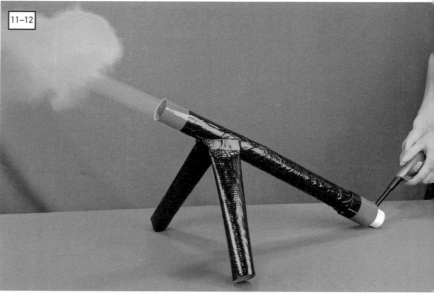

11 Quickly screw the cap back on, reposition your shot, and insert the BBQ lighter. It should fit snugly. If it's loose, the explosion may partially burst through the BBQ lighter opening.

Click the ignition and fire! You should hear a satisfyingly loud BANG! and see a small fireball (and ping-pong ball) explode from the barrel!

12 After each shot, unscrew the cap and blow air through the whole barrel. This clears out the gaseous by-product from the combustion reaction. If you don't clear the barrel, these gasses will prevent another combustion from occurring.

Pro Tip
If the ball falls through the barrel, poke two small holes near the middle, and insert a skewer or other thin, flame-resistant material through it. The skewer will prevent the ball from slipping past.

TROUBLESHOOTING
If the mortar isn't working well, there two things to check for:

First, the paper towel tube may be too big. The ping-pong ball should *just* be able to slide down the barrel. If there's an excess of space between the ball and the inside of the barrel, the explosion will travel *around* the ball. We want the explosion to remain contained *behind* the ball, thus forcing it outward.

You can decrease the size of the barrel's inner diameter by cutting another paper towel tube down one side, and inserting it into the barrel. Repeat until the ping-pong ball just barely slides down the barrel.

Second, there might not be the right mixture of fuel (hair spray) and oxygen. If you ignite the lighter and nothing happens, try this: Quickly remove the lighter, blow a gust of air into the ignition site, and reinsert the lighter to try again. Introducing a small bit of O_2 may be enough to get the combustion started.

If it still fails, then remove the ball, clear the barrel, and restart the firing process.

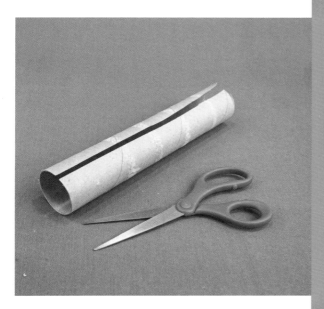

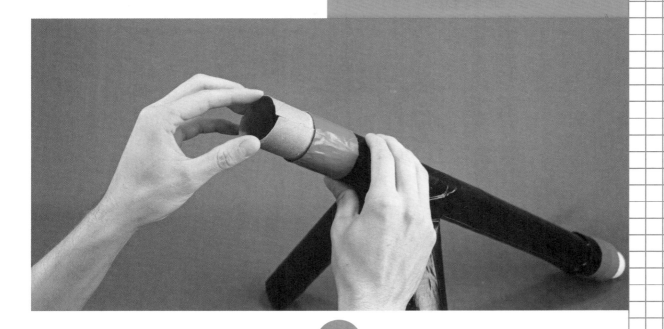

FIREARMS

117

SODA BOTTLE BOMBARD

Easily the loudest and most explosive project in the book, the soda bottle bombard is a force to be reckoned with! Using hair spray as fuel, and a BBQ lighter for ignition, you'll be able to fire a soda bottle missile at least 50' (15 m) in a long, graceful arc. Though relatively quick to build, it requires diligent construction and extreme safety to be enjoyed.

WARNING!!

This project is **EXCEPTIONALLY DANGEROUS!** You **must** use a 1-liter bottle or smaller, and you **must** use a bottle that's designed for carbonated beverages. If you use a larger bottle, the explosion will be too dangerous. If you use a bottle for noncarbonated beverages, it may explode. DO NOT use old or damaged bottles. You may use smaller bottles.

Additionally, this project is **EXTREMELY LOUD!** While testing this, my neighbors thought someone was firing a gun! Use ear protection, and fire only in areas where you won't disturb others.

TOOLS AND MATERIALS

- BBQ lighter
- 4" to 5" (10 to 13 cm) piece of ½" PVC pipe (depending on the length of your BBQ lighter)
- hot glue
- scissors
- utility knife
- two 1-liter carbonated beverage bottles
- craft foam
- duct tape
- tennis ball
- ear protection
- safety glasses
- gloves
- aerosol hair spray

MATERIAL SUBSTITUTION
Craft foam: You can use index cards or other thick paper. Craft foam is just more durable.

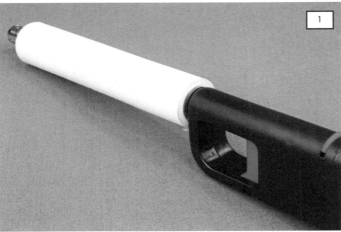

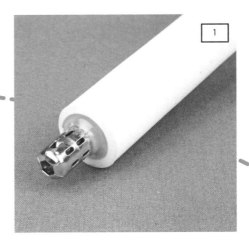

1 Create the bottle mount.
Slip the piece of PVC pipe over the BBQ lighter stem. Use liberal amounts of glue to seal the gap between the pipe and the lighter stem. Avoid getting any glue inside the tip of the lighter.

Important note: It's crucial to use *lots* of glue. If you don't, the hot gasses that are created during combustion will force their way through the gap between the pipe and the lighter stem, and possibly burn your hands.

FIREARMS

2. Use a utility knife to cut a small slit about 2" (5 cm) from the opening of one of the bottles, then use scissors to cut off the rest. Slip it over the pipe, and again, use lots of hot glue to hold it in place.

This important piece acts as a blast shield. Without it, the combustion that occurs will be blasted back toward your hands!

3. **Create the missile.**
Cut three identical fins out of craft foam, about 2½" × 1½" (6.4 × 4 cm) on the straight sides.

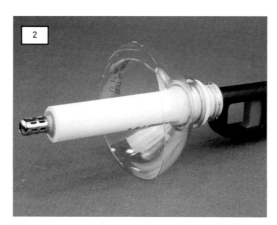

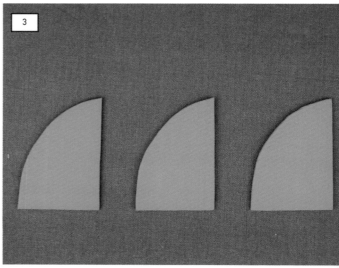

4. Use small pieces of duct tape to carefully attach both sides of each fin to the second bottle, near the opening. It's important that the fins are evenly spaced and attached as straight as possible.

5. Use the utility knife to carefully cut a tennis ball in half. Hot glue it onto the bottle, as shown.

This does three important things:

First, it adds a weight to the front of the missile, which helps stabilize the flight (see the note about leading weights on page 19).

Second, it makes it safer. The bottom of most bottles is quite rigid, so covering the bottom with a rubbery cushion reduces the chance of damaging something upon impact. The added weight also reduces the bottle's initial velocity, so it won't carry quite as much destructive energy when first fired, but it will increase its overall distance due to adding a leading weight.

Third, the rubbery cushion absorbs the impact upon landing, adding longevity to your missile.

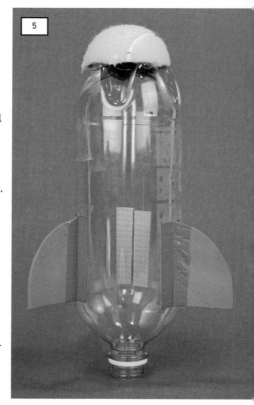

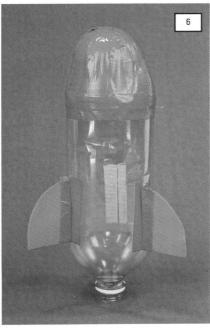

6 Rip at least twelve 8" × 1" (20 × 2.5 cm) strips of duct tape, and wrap them over the top of the tennis ball, as shown. This will prevent the tennis ball from breaking off upon impact (and it makes the missile look way better).

TROUBLESHOOTING
If the bombard doesn't fire, the #1 reason will be because it doesn't have the right ratio of fuel (hair spray) and oxygen. If you attempt to fire it and nothing happens, check to see if the flame from the lighter appears. If so, that means there's enough oxygen, but not enough fuel; add a quick burst of hair spray. If you don't see a flame, then there's too much fuel. Blow forcefully into the opening of the missile to introduce more O_2.

And, as mentioned, remove as much of the gaseous byproducts as you can after each launch.

7
Load and fire!
First, put on your safety glasses and ear plugs. For the first launch, you must wear gloves to protect your hands until you have tested the effectiveness of your hot gluing.

Fill the missile with a quick 1-second blast of hair spray. *Do not* overfill the bottle with hair spray: It won't work. You need some oxygen for combustion to occur.

8 Slip the missile onto the launcher. Aim very carefully, and pull the trigger! You will hear a resounding BANG! as the missile fires with an impressive amount of force.

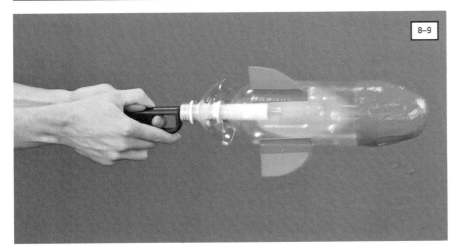

9
After each shot, the missile will be filled with gaseous by-products from the combustion reaction. You need to remove as much of these gasses as possible or it won't work a second time. Repeatedly squeeze the bottle and pop it back into its original shape to push out the gasses and fill it up with fresh air. You can also forcefully blow into the bottle a few times.

FIREARMS

HANDHELD ROCKET LAUNCHER

My personal favorite! The handheld rocket launcher takes all the power and excitement of a conventional PVC air cannon and packs it down into a sidearm-sized blaster! It takes under ten seconds to fully pressurize, and a simple valve modification allows for easy and quick release, perfect for achieving consistent, high-energy shots!

Tools and Materials

- at least 23" (59 cm) of schedule 40, ½" (1.3 cm)-wide PVC pipe
- measuring stick
- PVC cutters or saw
- PVC primer and solvent (glue)
- 1" (2.5 cm) metal clamp-in valve stem
- drill with ½" and ¼" bits (1.3 cm and 6.4 mm)
- utility cutters
- pliers
- 2" (5 cm) PVC plug
- 2" (5 cm) PVC coupling
- 2" to ½" (5 cm to 1.3 cm) PVC slip reducer
- ½" (1.3 cm) PVC ball valve
- ½" (1.3 cm) PVC tee
- two ½" (1.3 cm) PVC end caps
- spray paint, masking tape, and scrap paper (optional)
- hacksaw
- four 10" (25 cm) cable ties
- scissors
- 3½" (9 cm)-diameter hose clamp
- small air pump with hose and Schrader valve
- flathead screwdriver

MATERIAL SUBSTITUTIONS

Hose clamp: Duct tape

Small air pump: A regular-sized pump. (You may need to detach it between shots if it's too unwieldy.)

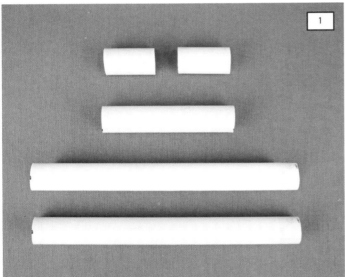

1 Cut the pipe into two 8" (20 cm), one 4" (10 cm), and two 1¾" (4.5 cm) pieces. If you're using a saw, clean up any PVC dust right away.

2 Drill a ½" (1.3 cm)-wide hole into the 2" (5 cm) PVC plug. Disassemble the valve stem.

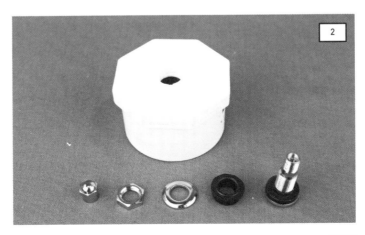

FIREARMS

3. Tire valves are designed to be inserted into a thin steel frame. The PVC plug is much thicker. You may need to carefully cut the valve's rubber washer in half with a utility knife to be able to install it in the next step.

4. Fit the valve stem through the hole in the PVC plug, then reassemble with the rubber washer, metal washer, and nut, in that order. Tighten the nut with pliers.

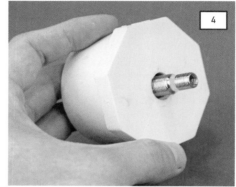

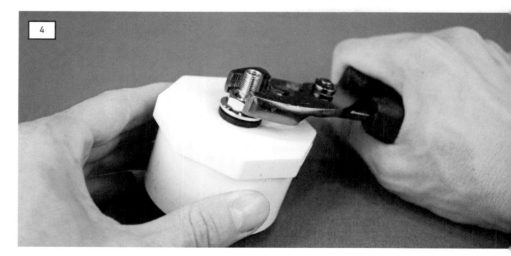

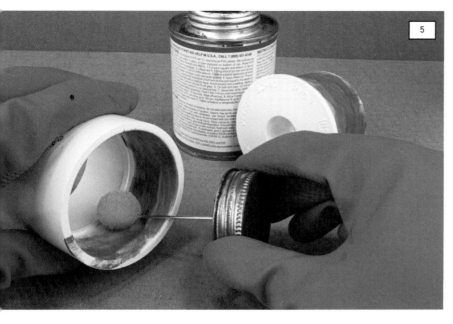
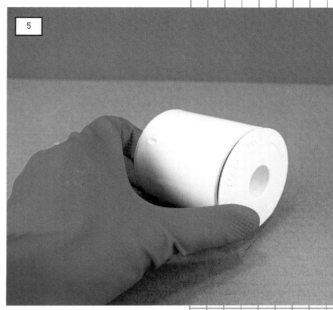

5 Begin gluing the PVC together. In this step, the 2" to ½" (5 cm to 1.3 cm) reducer is glued into the 2" (5 cm) coupling. Follow the instructions on the PVC primer and adhesive, wear gloves, work in a well-ventilated area, and cover your work surface with something disposable.

Typically, both parts that are being glued together will need a coat of primer and a coat of adhesive. Work quickly—the adhesive typically sets in about 30 seconds.

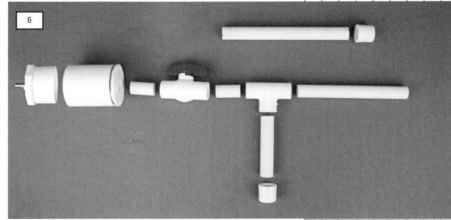

6 Continue assembling all the pieces in the arrangement shown here. Note the following:

- The plug with the valve stem is glued into the open side of the coupling.
- The short 1¾" (4.5 cm) pieces are used to connect the reducer to the valve, and the valve to the PVC tee.
- An end cap is glued to the bottom of the handle to prevent air from escaping.

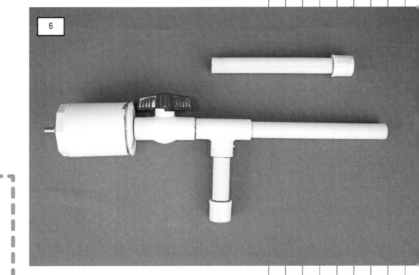

FIREARMS

7

(Optional) **Paint the PVC.**
A one-color coat of spray paint will add a lot of cosmetic value. You can also choose to mask off parts of the launcher to paint it in two colors, as shown here.

Once the first color of paint dries completely, mask off the painted part, and spray on your second color. Make sure to keep the valve stem covered to avoid clogging.

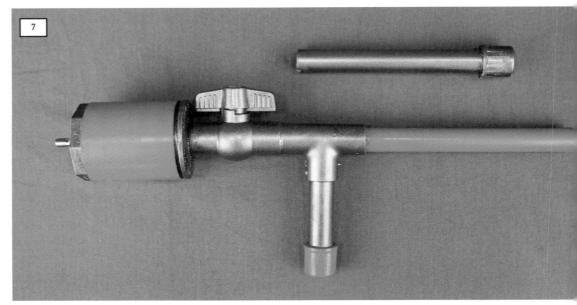

8

In this project, the key to a high-powered shot is to release the air pressure as rapidly as possible. This can be challenging because the built-in PVC valve is often quite stiff, and doesn't offer much leverage. To make it easier (and faster) to open the valve, you'll add a **valve handle extension**, which will give you more leverage.

Saw a notch out of the 8" (20 cm) pipe, starting about 1½" (4 cm) from the end. The angle of the notch is about 90 degrees, and cut about halfway into the pipe. The goal is to be able to fit this extension flush against the valve handle. Clean up the PVC debris as soon as you're done.

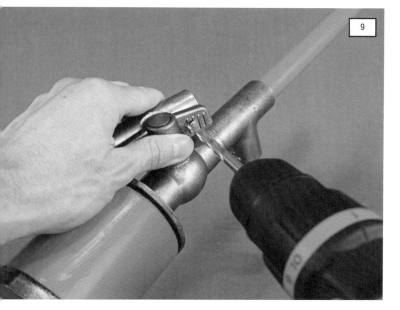

9

Drill a hole through the PVC valve handle and the handle extension. If you can pinch the two pieces together as you drill, you'll ensure that the holes are aligned. If not, drill through the PVC valve handle, and make a mark on the extension where the drill bit touches it. Then drill a hole into the extension separately.

10

Attach the extension onto the PVC valve with two cable ties, as shown. Strap them on as tightly as possible.

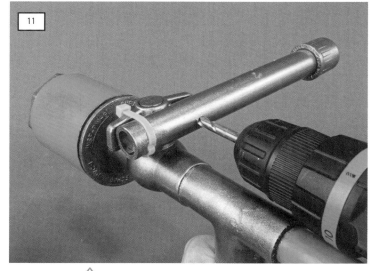

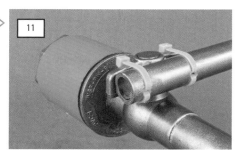

Repeat on the other side of the valve, and trim off the excess cable tie. You may need to drill through the pipe first this time.

FIREARMS 127

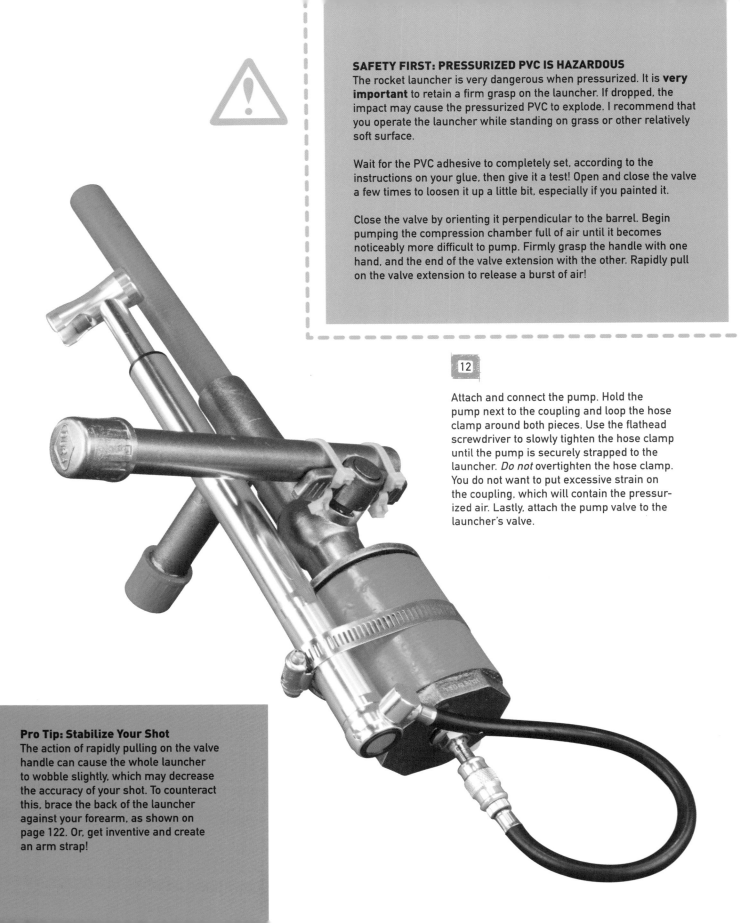

SAFETY FIRST: PRESSURIZED PVC IS HAZARDOUS

The rocket launcher is very dangerous when pressurized. It is **very important** to retain a firm grasp on the launcher. If dropped, the impact may cause the pressurized PVC to explode. I recommend that you operate the launcher while standing on grass or other relatively soft surface.

Wait for the PVC adhesive to completely set, according to the instructions on your glue, then give it a test! Open and close the valve a few times to loosen it up a little bit, especially if you painted it.

Close the valve by orienting it perpendicular to the barrel. Begin pumping the compression chamber full of air until it becomes noticeably more difficult to pump. Firmly grasp the handle with one hand, and the end of the valve extension with the other. Rapidly pull on the valve extension to release a burst of air!

12

Attach and connect the pump. Hold the pump next to the coupling and loop the hose clamp around both pieces. Use the flathead screwdriver to slowly tighten the hose clamp until the pump is securely strapped to the launcher. *Do not* overtighten the hose clamp. You do not want to put excessive strain on the coupling, which will contain the pressurized air. Lastly, attach the pump valve to the launcher's valve.

Pro Tip: Stabilize Your Shot
The action of rapidly pulling on the valve handle can cause the whole launcher to wobble slightly, which may decrease the accuracy of your shot. To counteract this, brace the back of the launcher against your forearm, as shown on page 122. Or, get inventive and create an arm strap!

CREATE THE ROCKETS

Many things can be fired from the rocket launcher: Mini marshmallows, toy gun foam darts, and potato chunks are all good choices. However, in my opinion, the best projectile is a paper and duct tape rocket! They're accurate, far-flying, and durable!

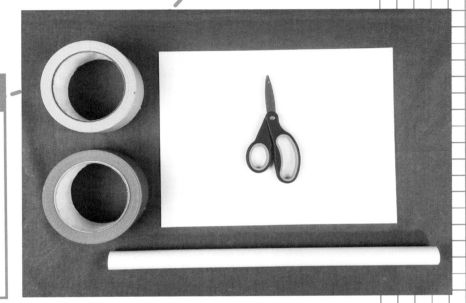

Tools and Materials

- 4 paper towel tubes
- extra piece of ½" (1.3 cm) PVC pipe, at least 12" (30 cm) long
- duct tape
- 8½" × 11" (21.6 × 28 cm) card stock*
- scissors

*Regular copy paper can be substituted for card stock, however it's more difficult to get straight fins.

Wrap a layer of duct tape around the PVC pipe. This will add just enough width to the inner diameter of your rocket to ensure it slides easily over the barrel. This is especially important if you painted it.

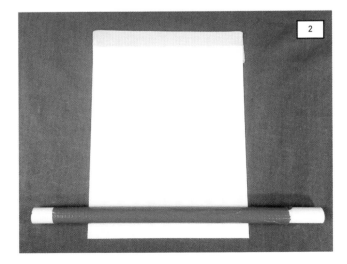

2
Form the tube. Apply a piece of duct tape along the 8½" (22 cm) side of the paper, with about half of the adhesive exposed, as shown. Lay the pipe across the other side.

FIREARMS

3

Tightly roll up the paper around the pipe. As you finish, use the piece of duct tape from the last step to seal the paper tube closed.

4

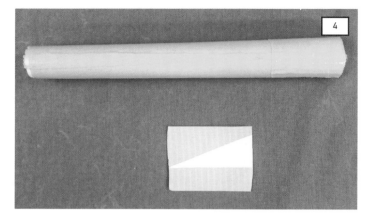

Cover the tube with more duct tape and begin making the fins. Cut a rectangle that's about 2½" (6.4 cm) long and 1" (2.5 cm) wide, then cut that diagonally to make two triangles. Apply a piece of duct tape to one side of the triangle, as shown.

5

Carefully attach the bottom of the fin near the back of the fuselage, as shown. Make sure that the fin is straight

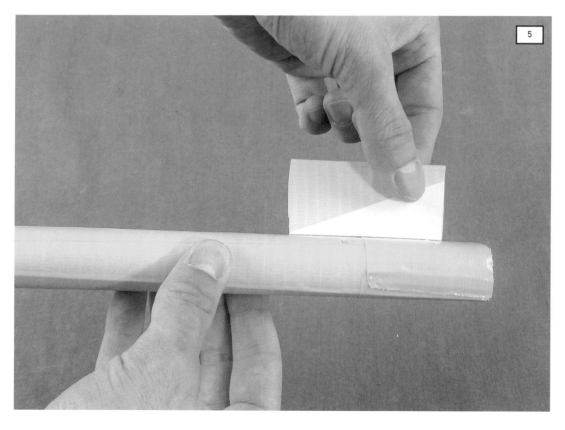

6 Fold the remaining tape over the fin and cover the rest with a small piece of tape. Covering the fins in tape will increase their durability.

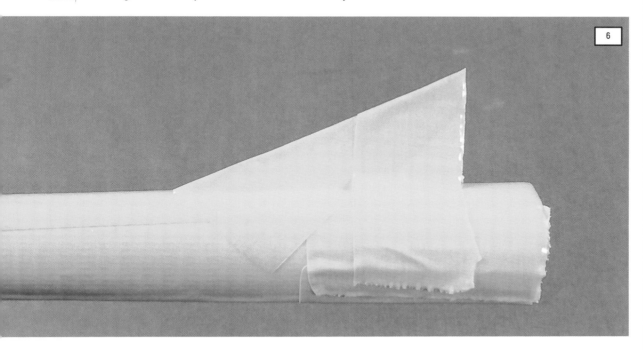

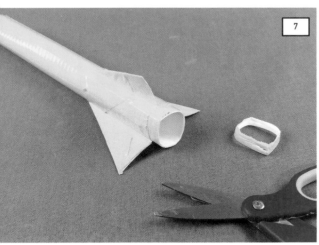

7 Repeat until you have a total of three fins. Cut off the back of the rocket fuselage to create a clean edge that will fit easily over the barrel.

8 Build the nose cone (this is very similar to the blow gun darts on pages 13 to 14). Cut a square of card stock that's about 3½" × 3½" (9 × 9 cm), then cut a slit from one side to the center.

FIREARMS   131

9

Overlap one side of the paper, as shown, creating a cone shape. Continue wrapping one side of the paper around the cone shape until you have a narrowly-tapered point. Apply a small piece of tape to hold the cone together.

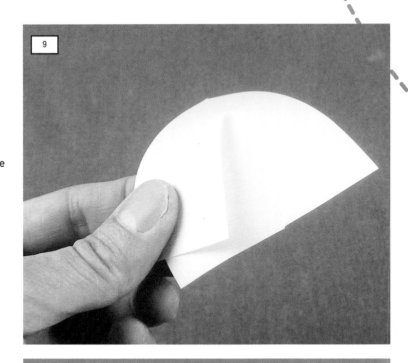

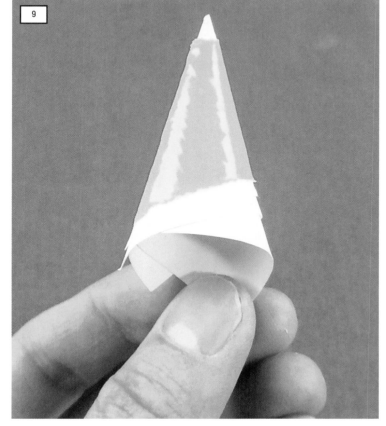

11

Attach the cone to the tip of the rocket with several thin lengths of tape that run from the tip of the nose cone to about 3" (7 cm) down the length of the rocket. You're done!

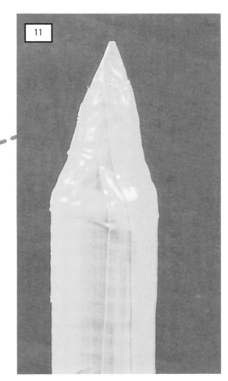

10

Cut off the bottom of the cone so the edges are flush with the tabletop as shown. The diameter of the bottom of the cone should be about as wide as your rocket tube.

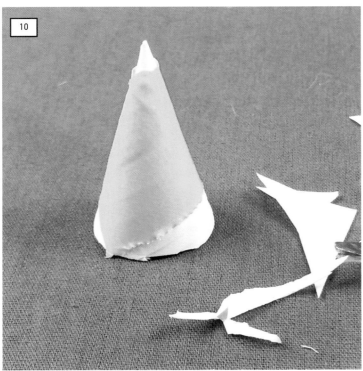

BALLISTICS GEL TARGET

So you've fired your crossbow into a cardboard target, hit cans with rubber bands, and battered your bed pillows with a ballista. Now what? Get serious with your target practice by mixing up a batch of homemade ballistics gelatin! This recipe creates a dense gel that will let you compare the strength of your DIY shooters by seeing how far the projectiles penetrate it.

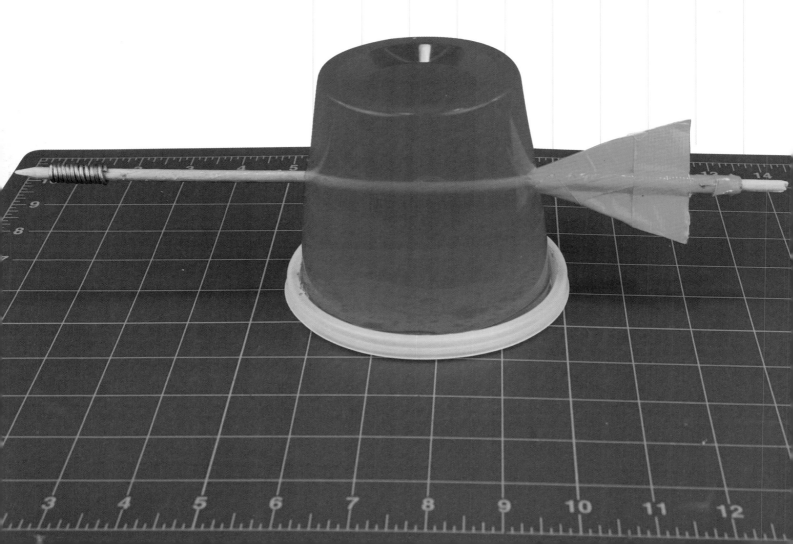

TOOLS AND MATERIALS

- container for molding the gel
- 1-cup (240 ml) measuring cup
- saucepan
- water
- unflavored gelatin
- whisk
- cooking oil spray (optional)

Decide on the gel shape. Find a container with a smooth, nonporous surface, preferably one that's bowl shaped. (If the container has straight sides, it might be hard to remove the gel.)

Using the measuring cup, fill the container with room-temperature tap water. Keep track of how many cups of water you add. For each 1 cup (240 ml) of water, you'll need 1 oz (30 g) of gelatin. For this example, I'm making a small 3-cup (720 ml) batch of gel.

FIREARMS

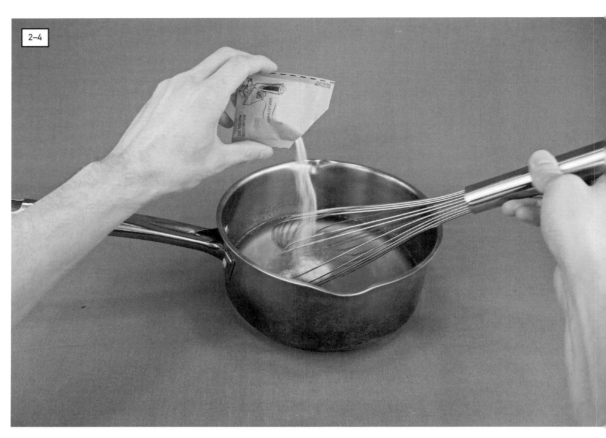

2

Pour the water from the container into a saucepan. Whisk in one packet of gelatin at a time to minimize the amount of clumping.

3

Put the pan on the stove over medium-high heat. Once the water-gelatin mixture starts to heat up, reduce the heat to low—you do not want the mixture to boil. Stir continually and gently (to avoid bubbles) until the gelatin is fully dissolved. Turn off the heat.

4

(Optional) Prepare the mold—it's your choice. Spraying the mold with cooking oil will make it easier to remove the gel. However, it will also give the gel a greasy surface (which you can blot off with paper towels), and a mottled texture, which makes it less transparent. For the example shown here, I chose not to use the spray.

5

Pour the gel mixture into the mold and refrigerate. Depending on the size of the mold, this can take anywhere from 2 to 8 hours.

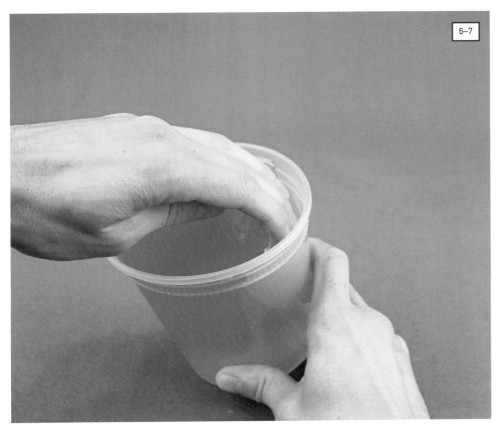

Remove the gel from the mold: If you didn't use cooking oil, pry the gel loose by gently slipping your fingers between the gel and the mold. Once it's been loosened all around, turn the mold upside down and tap it firmly on a sturdy surface.

Get ready to test!
Place your gel in an area where it's safe to fire—to make sure your shot won't ricochet back at you (in the unlikely event that you miss). Because the gel is highly absorbent, you can fire at point-blank range. Otherwise, take careful aim, and fire! Fire different contraptions into the gel in different places to measure the power of each one!

Modify the gel density
This ratio of 1 ounce (30 g) of gelatin to 1 cup (240 ml) of water creates a very dense gel, which is suitable for the higher-powered contraptions in this book, such as the PVC Bow. For a less-dense mixture that's better suited for things like the BBQ Blaster, try using a ratio of ¾ ounce (22 g) to 1 cup (240 ml) of water. For a Jell-O–like consistency, you can go as low as ½ ounce (15 g) per 1 cup (240 ml) water.

FIREARMS

RESOURCES

I designed each project in this book to be as accessible as possible. Everything is made from office supplies and/or common materials from hardware stores. The few things that are a little more difficult to get hold of, such as a metal bicycle pump, I included because the payoff is so worth it. If your hardware store doesn't have it, you can find it online.

ACKNOWLEDGMENTS

Heartfelt thanks go out to my parents, Nancy and Jim, and my colleagues at Galileo Learning for being endlessly supportive. Another shout-out to my sister Kaile for shooting some difficult-but-totally-worth-it action shots for this book. And to Ali, for putting up with months of explosive noises from the garage and not asking too many questions when I come inside smelling like burnt hair spray.

ABOUT THE AUTHOR

Lance Akiyama is a maker, writer, and educator. He creates free-to-use engineering projects for kids under the moniker Lance Makes online on Instructables and Patreon. He's also currently employed as a STEAM-based curriculum developer for Galileo Learning. Lance spends his free time making plans to survive the zombie apocalypse, browsing craft stores for the latest googly eyes, and attempting to fix up his house using only cardboard and duct tape. He resides in the San Francisco Bay Area. To get in touch, contact him at LanceMakes.com.

INDEX

A

aerosol hair spray
    Ping-Pong Ball Mortar, 113
    Soda Bottle Bombard, 119
air pump, for Handheld Rocket Launcher, 123

B

Ballista
    bipod, 77
    bolts, 78
    firing, 79
    frame, 73–74
    introduction to, 72
    loading, 79
    materials, 73
    storing, 79
    stringing, 76–77
    throwing arms, 74–76
    tools, 73
    torsion bundle, 74–76
    trigger, 77
Ballistics Gel Target
    density modification, 137
    introduction to, 134
    materials, 135
    testing, 137
    tools, 135
bamboo skewers
    Ballista, 73
    Onager, 59
BBQ Blaster
    barrel, 107–108
    BB Blaster, 111
    firing, 111
    fuel line, 110
    ignition installation, 110
    introduction to, 106
    lighter modification, 108
    loading, 111
    materials, 107
    resetting, 111
    safety, 111
    tools, 107
    wiring test, 109
BBQ lighters
    BBQ Blaster, 107
    Ping-Pong Ball Mortar, 113
    Soda Bottle Bombard, 119
BBs, for BBQ Blaster, 107
binder clips
    Ballista, 73
    Desk Drawer Booby Trap, 89
    Duct Tape & PVC Crossbow, 45
    Micropult, 17
    Slingshot and Arrow, 39
bottle caps
    Da Vinci Catapult, 67
    Onager, 59
bows & slingshots
    Duct Tape & PVC Crossbow, 44–49
    Pulley-Powered PVC Bow & Arrow, 28–37
    Slingshot and Arrow, 38–43
    Wrist-Mounted Crossbow, 50–55

C

cable ties
    BBQ Blaster, 107
    Handheld Rocket Launcher, 123
clothespins
    Roller-Amplified Many-Thing Shooter, 83
    Wrist-Mounted Crossbow, 51
cork
    Da Vinci Catapult, 67
    Poker Chip Shooter, 95
    Roller-Amplified Many-Thing Shooter, 83
    Wrist-Mounted Crossbow, 51
craft foam
    Roller-Amplified Many-Thing Shooter, 83
    Soda Bottle Bombard, 119
    Wrist-Mounted Crossbow, 51

craft sticks
- Ballista, 73
- Da Vinci Catapult, 67
- Micropult, 17
- Onager, 59
- Poker Chip Shooter, 95
- Rubber Band Rockets, 19
- Slide-Action Rubber Band Gun, 101
- Slingshot and Arrow, 39
- Wrist-Mounted Crossbow, 51

curious contraptions
- Desk Drawer Booby Trap, 88–93
- Poker Chip Shooter, 94–99
- Roller-Amplified Many-Thing Shooter, 82–87
- Slide-Action Rubber Band Gun, 100–103

D

Da Vinci Catapult
- catapult base, 68
- craft sticks, bending, 67
- firing, 71
- introduction to, 66
- loading, 71
- materials, 67
- power source, 69–71
- supports, 68
- tools, 67
- winding drum, 69

dental floss, for Ballista, 73

Desk Drawer Booby Trap
- arming, 92
- catapult, 89–90
- introduction to, 88
- materials, 89
- tools, 89
- trigger, creating, 90–91
- trigger setup, 92–93
- trigger, testing, 91

dowels
- Da Vinci Catapult, 67
- Duct Tape & PVC Crossbow, 45
- Onager, 59
- Poker Chip Shooter, 95
- Pulley-Powered PVC Bow & Arrow, 29
- Roller-Amplified Many-Thing Shooter, 83
- Slingshot and Arrow, 39

Duct Tape & PVC Crossbow
- bolt clip, 48
- bolts, 48–49
- bow, attaching, 46
- bow, stringing, 46
- firing, 49
- introduction to, 44
- loading, 49
- materials, 45
- safety, 44, 49
- stock assembly, 45–46
- tools, 45
- trigger, 47–48
- trigger notch, 47

F

firearms
- Ballistics Gel Target, 134–137
- BBQ Blaster, 106–111
- Handheld Rocket Launcher, 122–133
- Ping-Pong Ball Mortar, 112–117
- Soda Bottle Bombard, 118–121

floral wire
- Pulley-Powered PVC Bow & Arrow, 29
- Slingshot and Arrow, 39

G

gelatin, for Ballistics Gel Target, 135

H

Handheld Rocket Launcher
- assembly, 123–128
- introduction to, 122
- materials, 123, 129
- painting, 126
- rockets, 129–133
- safety, 128
- shot stabilization, 128
- tools, 123, 129
- valve handle extension, 126

hose clamp, for Handheld Rocket Launcher, 123

I

index fletching, 35

M

magnet, for BBQ Blaster, 107
mason line, for Pulley-Powered PVC Bow & Arrow, 29
medicine bottle, for BBQ Blaster, 107

Micropult
- fins, 19
- firing, 17
- fuselage, 19
- introduction to, 16
- leading weights, 19
- materials, 17
- projectile basket, 17
- tools, 17

mini medieval siege machines
- Ballista, 72–79
- Da Vinci Catapult, 66–71
- Onager, 58–65

N

nails, for Straw Blowgun, 13

O

Onager
- base, 59
- firing, 65
- introduction to, 58
- materials, 59
- storing, 65
- throwing arm, creating, 60–62
- throwing arm, installing, 62–63
- tools, 59
- trajectory adjustment, 65
- trigger, creating, 63–64
- trigger, testing, 64
- upright, 59–60

P

paint stirrers
- Da Vinci Catapult, 67
- Desk Drawer Booby Trap, 89
- Poker Chip Shooter, 95
- Roller-Amplified Many-Thing Shooter, 83
- Slide-Action Rubber Band Gun, 101
- Wrist-Mounted Crossbow, 51

paper clips
- Ballista, 73
- BBQ Blaster, 107
- Da Vinci Catapult, 67
- Desk Drawer Booby Trap, 89
- Micropult, 17
- Onager, 59
- Straw Blowgun, 13

paper towel tubes
- Handheld Rocket Launcher, 129
- Ping-Pong Ball Mortar, 113

paracord, for Duct Tape & PVC Crossbow, 45

pens
- BBQ Blaster, 107
- Pocket Bow, 23
- Roller-Amplified Many-Thing Shooter, 83
- Rubber Band Rockets, 19

Ping-Pong Ball Mortar
- barrel, 113–114
- bipod, 115
- firing, 116
- ignition site, 115
- introduction to, 112
- loading, 116
- materials, 113
- tools, 113
- troubleshooting, 117

plastic bottles
- Ping-Pong Ball Mortar, 113
- Soda Bottle Bombard, 119

Pocket Bow
- arrow, 24
- bows, 23–24

firing, 25
introduction to, 22
materials, 23
tools, 23
Poker Chip Shooter
    ammo container, 97–98
    firing, 99
    firing mechanism, 98
    foundation, 96–97
    handle, 99
    introduction to, 94
    loading, 99
    materials, 95
    precision and, 95
    safety, 99
    tools, 95
    troubleshooting, 99
Pulley-Powered PVC Bow & Arrow
    arrow rest, 36
    arrows, 34–35
    bow, creating, 29–31
    bow, stringing, 32–33
    firing, 37
    firing string, 36
    hand dominance, 36
    index fletching, 35
    introduction to, 28
    materials, 29
    safety, 28, 37
    storing, 37
    tools, 29
    troubleshooting, 37
PVC pipe
    Duct Tape & PVC Crossbow, 45
    Handheld Rocket Launcher, 123, 129
    Pulley-Powered PVC Bow & Arrow, 29
    Soda Bottle Bombard, 119

R
Roller-Amplified Many-Thing Shooter
    frame, 83–84
    introduction to, 82
    materials, 83
    roller, attaching, 84–86
    rubber band, attaching, 84–86
    shot preparation, 87
    tools, 83
Rubber Band Rockets
    firing, 20
    fuselage, 19
    introduction to, 18
    leading weights, 19
    materials, 19
    safety, 18
    tools, 19
rubber bands
    Desk Drawer Booby Trap, 89

Pocket Bow, 23
Poker Chip Shooter, 95
Roller-Amplified Many-Thing Shooter, 83
Rubber Band Rockets, 19
Slide-Action Rubber Band Gun, 101
Slingshot and Arrow, 39
Wrist-Mounted Crossbow, 51
rulers
    Desk Drawer Booby Trap, 89
    Duct Tape & PVC Crossbow, 45
    Pulley-Powered PVC Bow & Arrow, 29

S
safety
    BBQ Blaster, 111
    disclaimer, 9
    Duct Tape & PVC Crossbow, 44, 49
    Handheld Rocket Launcher, 128
    Poker Chip Shooter, 99
    Pulley-Powered PVC Bow & Arrow, 28, 37
    Rubber Band Rockets, 18
    Slingshot and Arrow, 38
    Soda Bottle Bombard, 118, 119
simple & successful
    Micropult, 16–17
    Pocket Bow, 22–25
    Rubber Band Rockets, 18–21
    Straw Blowgun, 12–15
Slide-Action Rubber Band Gun
    design variables, 103
    firing, 103
    handle, 102
    introduction to, 100
    loading, 103
    loading notches, 101
    materials, 101
    measurements, 102
    slide-stops, 103
    tools, 101
    trigger, 102
    troubleshooting, 103
Slingshot and Arrow
    arrow rest, 41–42
    arrows, 42
    firepower, 42
    firing, 43
    introduction to, 38
    loading, 43
    materials, 39
    safety, 38
    slingshot frame, 39–40
    slingshot supports, 40–41
    tools, 39
Soda Bottle Bombard
    bottle mount, 119–120
    firing, 121
    introduction to, 118

loading, 121
materials, 119
missile, 120–121
safety, 118, 119
tools, 119
troubleshooting, 121
Straw Blowgun
    blow tube, 13
    darts, 13–14
    firing, 15
    introduction to, 12
    materials, 13
    tools, 13
straws
    Pocket Bow, 23
    Pulley-Powered PVC Bow & Arrow, 29
    Slingshot and Arrow, 39
    Straw Blowgun, 13
    Wrist-Mounted Crossbow, 51
string
    Da Vinci Catapult, 67
    Desk Drawer Booby Trap, 89
    Onager, 59
    Pulley-Powered PVC Bow & Arrow, 29
    Wrist-Mounted Crossbow, 51

T
tennis ball, for Soda Bottle Bombard, 119
troubleshooting
    Ping-Pong Ball Mortar, 117
    Poker Chip Shooter, 99
    Pulley-Powered PVC Bow & Arrow, 37
    Slide-Action Rubber Band Gun, 103
    Soda Bottle Bombard, 121
    Wrist-Mounted Crossbow, 55

V
valve stem, for Handheld Rocket Launcher, 123

W
Wrist-Mounted Crossbow
    bolts, 55
    crossbow, creating, 51–52
    crossbow, stringing, 52
    firing, 55
    introduction to, 50
    loading, 55
    materials, 51
    padding, 54
    straps, 54
    tools, 51
    trigger, 52–53
    trigger handle, 53
    troubleshooting, 55

[ALSO AVAILABLE]

Duct Tape Engineer
978-1-63159-130-3

Rubber Band Engineer
978-1-63159-104-4

PVC and Pipe Engineer
978-1-63159-334-5

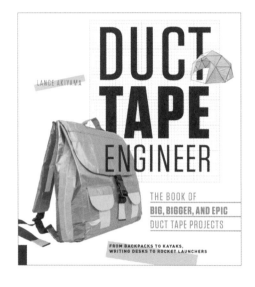

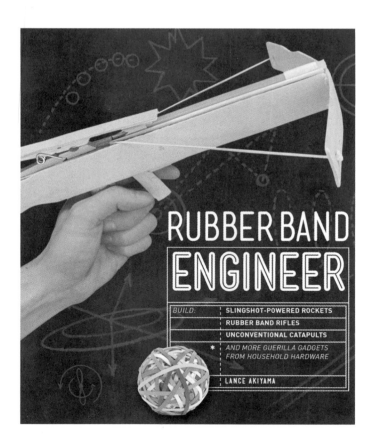

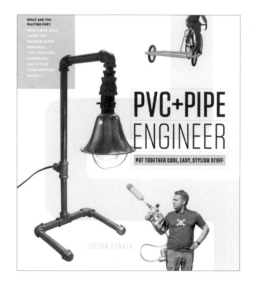